Scotland's Landscape

Scotland's Landscape
Endangered Icon

Anna Paterson

Polygon at Edinburgh
An imprint of Edinburgh University Press Ltd
22 George Square, Edinburgh

Typeset in Minion by
Pioneer Associates, Perthshire, and
printed and bound in Great Britain by
The Cromwell Press, Trowbridge, Wiltshire

A CIP Record for this book is available from the British Library

ISBN 0 7486 6272 3 (paperback)

Contents

	List of Illustrations	vi
	Preface	vii
	From Ekman: The Forest of Hours	x
	Acknowledgements	xii
1	The Scottish Environment in Fiction and Fact	1
2	Literature and Perceptions of National Identity	17
3	The Official Environment	49
4	Education and the Environment	77
5	Stewardship of the Land I: Scotland's Landscape – the Resource	94
6	Stewardship of the Land II: Conserving or Marketing?	121
7	Greening the Mean Streets?	157
8	The Environmental Economy: Research and Development, Local Trade and Big Business	191
9	Research and Development in the Countryside	213
10	What Is Left of Scotland's Landscape?	244
	Bibliography	258
	Index	266

List of Illustrations

1. The banks of Loch Morlich 14
2. Deerstalkers on the Auchlyne Estate 22
3. The Kvaerner shipyard at Govan 25
4. Crofting at Melness, Sutherland 64
5. The West Highland Way, north of Milngavie 81
6. Ascending Beinn a'Bhuird 99
7. The McLaren Community Leisure Centre in Callander 110
8. Balloch marina on Loch Lomond 127
9. Scots pine trees on the Mar Estate 133
10. The Kyle of Tongue 152
11. Vertical Garden City in Datatown 160
12. Leslie Court in Fairfield, Perth 169
13. Model for a centre for outsider art 175
14. New Lanark industrial heritage village 186
15. Windfarm in Datatown 197
16. Wooden motorway overhead display 207
17. Multi-storey Forest in Datatown 224
18. A block of planted conifer in the Cairngorms 233
19. A fish farm in Argyll 246
20. Peat cutters on Bernera, Lewis 252
21. Jawbones of culled deer on the Mar Estate 255

Preface

SCOTLAND'S LANDSCAPES and cityscapes have become emblematic of the nation, parts of the citizens' sense of belonging – icons, to be venerated even if not understood. People have taken in the thousands of images and gone to see for themselves: the high hills and rolling heaths, the lochs and 'Queen's Views', the castles and villages and towns built of old grey blocks of stone. They are elements of the Scottish sense of identity, but it does not take long to realise that the natives take a quite ruthless view of the 'environmental' aspects of their countryside and their towns. I use the concept 'environment' in the common but perhaps questionable sense of meaning something special, a setting that deteriorates unless cared for knowingly and systematically: fresh air and water can become polluted, nature impoverished and natural resources exhausted, and the 'built environment' so inhuman that it causes illness and disorientation.

By narrowing the focus to Scotland, and then again to only some of the many strands in Scottish public life, I have deliberately stayed with a limited, local version of a global process. The original reason was that the Scottish themselves seemed a special case of national ambivalence, unusually unwilling to see the need to bridge the gap between landscape image and environmental reality.

It turned out that there are signs that the newly devolved Scotland could also become part of a global process, identified by radical economists as 'Corporate Takeover', 'Globalisation', 'Death of Democracy' and so on. The elements include a majority of apolitical citizens, a weak and therefore potentially corruptible political establishment, and a society with a growing dependence on money handouts from very large and very, very wealthy supranational business interests. Precious values seem to have less and less chance of surviving in the new corporation-dominated world. They include real democracy (as opposed to the uncommitted 'consensual partnership' variety), justice, freedom of expression, civic and environmental care. While politics might fail, people's attention is at least being attracted by insistent whistleblowing. It would be pleasing to believe that in a small way, I could contribute to these shrill warning noises. Scotland's much-loved landscapes and cityscapes are as good a cause to shout about as any.

I have written this book as an ordinary citizen wanting to find out 'what's going on'. I have had no academic or journalistic framework for support, no research assistants and no grants. To a large extent, I have relied on asking knowledgeable people questions and I am deeply grateful to all those who have talked to me. Throughout the book, there are summarised notes from these interviews, which were more about listening to views than following a special line of inquiry. The list of 'interviewees' comes in the Acknowledgements section on page xii below. Although I have tried to get the acceptance of the person talking for the formal summaries, I alone must take the responsibility for any misunderstanding or mistakes.

Rooting around among the written evidence available to the public, I have come across many strange and contradictory things that require investigation on a grander scale than my time allowed. I have become convinced that there is real cause for concern. The people who decide what happens to the environment in Scotland are not primarily its elected representatives, local or national. Still, although the overall picture remains one of resource-depletion for immediate gain, the flow of official reassurance has been accompanied by some truly useful work by organisations and individuals.

Among the people who are important to the understanding of Scotland's environment are Professor T. C. Smout, Environmental Historian and Historiographer Royal of Scotland, and Kevin Dunion, Director of Friends of the Earth Scotland and member of the Scottish Executive's Ministerial Group on Sustainability. I am very grateful for their contributions. I also want to acknowledge a huge debt of gratitude to Kerstin Ekman, not only one of Sweden's greatest writers, but the most subtle of 'environmentalists'. My family and friends have been kind, patient and helpful; Alan and Andrew deserve to be singled out for being supportive in every way. As were the staff at Polygon at Edinburgh, to whom – and especially to Nicola Carr – I am for ever grateful.

I am also grateful to all the people who have let me use their images to illustrate this book, but especially to Colin McPherson, who has been a source of interesting and entertaining talk, and has provided endless images of Scotland as she is.

The only image I would have liked to add is Jeff Wall's great picture *A Gust of Wind*.* It shows an exhausted landscape with the crowded blocks of a cityscape in the distance. It is a beaten landscape, but one where a gust of wind is hitting back at the people who made it by scattering their bundles of documents. The bleakness that surrounds them is the heritage of those who do not care and do not see.

**A Gust of Wind (After Hokosai)* by Jeff Wall is on display at Tate Modern, London

From Ekman: The Forest of Hours

STRANDS OF TIME run through the forest. The high fields of scree are solidified waves of stone, long swells of unmoving time. Tall trees, once whispering in the wind, have sunk into the peatbogs, where the marshy pools are fermenting a brew of time. Here and there, flowering woodland penetrates the darkness of the firs and the sea of stones, forming wedges of broad-leaved trees, fragrant night-flowering plants and humming frail-winged insects. There, the noble trees are singing. The leaves of linden and hazel are dancing in a gentler wind and their roots send fibrils into a richer soil than the meagre ground under the firs. It is forgotten woodland, flowering in borrowed time . . .

People have always tried to slash and burn their way into the forest. And when it proved resilient, when even the goats failed to tear the bark off the trees, they tried to control it by naming. They found everything malevolent in the forest: stinkweed and wormwood, devil's nettle, dead man's bells and poison parsley. They named the plants which neither stank nor stung nor killed according to who fed on them: hawkweed, bear's foot, cowgrass, swine's snout and chickweed, hare's thistle, bee's nest and cuckoo's meat. For some, playfulness and for others plain practicality.

The rest became known as grass to most people, but there were those who named also the useless and the lovely ones: sweetfern and rose bay willow, forget-me-not, sweet cicely, water nymph, angel's eye and herb of grace, such were their names.

But the forest just grew and flowered. It flowered with senseless frenzy, remote from their names, and some wanderers came all the way, down into the deepest hollows under the stones and the darkness of marshland pools, and onto the steep silent braes of the lochs and they named the alien growths in these places and called them bog onion and brittle bladder fern, ghost orchid and pinesap. But it was not enough. The forest flowered on heedlessly, long after they had been silenced, roots twisting through their gaping mouths. It flowered known and named only by those who hummed and clicked and twittered, by the rustle of wings and rattle of claws and thud of antlers against tree trunks.

(*The Forest of Hours* by Kerstin Ekman, trnsl. Anna Paterson, Chatto & Windus, 1998 pp. 310–11)

Acknowledgements

THE FOLLOWING people talked to me about aspects of Scotland's environment. The first group is quoted by name in various chapters. The second group has also contributed enormously by talking and providing information, helping and advising. I am very grateful to them all, including those who provided me with unattributable information.

Roger Crofts, James Dalhousie, Kevin Dunion, Tasso and Margaret Elephtheriou, Geoff Fagan, Arne Jernelöv, Sandy Halliday, Robin Harper, Syd House, Alison Kennedy, Howard Liddell, Thomas Lindhquist, Rory MacLellan, Fraser Middleton, Mats Olsson, Michael Osborne, James Robertson, Christopher Smout, John Smyth, Mark Wells, Colin Wishart, Gunnar Zettersten.

Bengt-Erik Bengtsson, Ian Blyth, Hans Borg, Janet Brand, Geri Cruickshank, Alison Glenn, John Gunn, Veilla Hill, Dag Klockby, Alan and Andrew Paterson, Rebecca Tarschys, Cecilia Ruben, Deyan Sudjic, Andrew Whitehouse.

1

The Scottish Environment in Fiction and Fact

Scotland's Landscape: Icon or Environment?

LANDSCAPE IMAGERY is part of daily life in Scotland. Glens and lochs and seashores have combined into an iconic national landscape, displayed everywhere with pride and affection. Only Scotland's history – or at least its more dramatic moments – appears to be as powerful a source of the Scottish sense of identity. Both historical episodes and landscape scenes are used in a national marketing exercise, seemingly as effective with local people as with foreigners. The potency of the landscape images is confirmed by their capacity to shift commodities. Scottish nature looks appealing enough to persuade people to buy almost anything, from shortbread to inward investment options.

It is so much more surprising then that more practical and tangible signs of identifying with nature are so weak. All indicators seem to point the same way: compared with many other western European people, the Scottish have shown little interest either in nature conservation or in the many issues grouped under the general heading of 'environmentalism'. It is this disconnection between image and reality, and between human society and the landscape 'outside' that is the subject matter of this book. This quote from Alasdair Gray's remarkable novel *Lanark*[1] sums up the result of my researches:

> One morning he said to the nurses making the bed, 'What's outside the window?'
>
> 'Just scenery. Miles and miles of scenery.'
>
> ' Why are the blinds never raised?'

'You couldn't stand the view, Bushybrows. We can't stand it and we're perfectly fit.'

I want to follow up the way nature is treated both in Scottish culture and in policy, on the assumption that in a democratic state, policy is at least symptomatic of the cultural beliefs of the citizens. Because of Scotland's present state of political and cultural transition, it also seemed important to add interviews with key people to the reading of background material. The project looked ready to start ramifying endlessly unless some principles of selection were introduced.

Starting with culture, I felt it was necessary to focus on text. Although the visual arts, films, TV programmes and so on are vital evidence of how a society goes about its business, writing is still a dominant means of expression. So I have looked in books and magazines and newspapers for clues of how the nation feels about the environment, and also in the writing by public bodies like the Scottish Office Executive, quangos[2] and voluntary organisations. One of the most fascinating parts of the reading was trying to find the Scottish landscape in recent works of fiction. From the early 1970s onwards, literary writing has been a flagship of Scottish national consciousness and cultural renewal, and arguably a driving force behind the campaign for increased political independence in 1979. Reading as many books as possible from the last thirty years, I looked for how 'nature' was written about, if at all. The rarity of novels that integrated the non-human world into the existence of the human characters was striking.

Academic writing is important too, but emanates from a complex, rule-bound world and has to be 'translated' for the general public into simplified messages. Until the last decade or so, there have been few investigations of the role of the environment in Scotland outside the framework of the natural sciences or industrial research. Scottish environmental history is one of the first fields of study suggesting a change of heart.[3] Although the social sciences such as social anthropology have been examining people/nature relationships, these studies are relatively rarely set in Scotland. Cultural studies is another relevant area that has grown throughout the 1990s. Work on national identity and the heritage industry in Scotland has taken off, influenced by books such as *Scotland – The Brand* by David McCrone and his colleagues.[4]

Politics is another form of cultural activity that should directly reflect the concerns of the citizens. 'The environment' has not ever come close to being a serious issue in Scottish politics. Part of the reason might be that until recently, the politicians have been working too far away from home on a London stage, but they are also responding to a widespread lack of interest in their home territory. In the new political world presided over by the devolved Parliament, Scotland's iconic landscapes have featured quite widely in the discourse, but there has been little evidence of any political will to generate new policies with dominant environmental content.

To get a better grasp of what environmental awareness has meant for Scottish society, three chapters are devoted to examining some practical outcomes. These have been roughly subdivided as relevant to respectively the countryside, the 'built landscape' and to some aspects of the 'green economy'. The idea is not so much to provide complete accounts and complete prescriptions, but to look at how rhetoric and reality match. In each case, there will be quotes from what has been written and/or said by professional practitioners about Scottish problems in the selected areas. In this group of chapters, as in Chapter 2, the hardest task was selection of certain areas and elimination of others. Some of the reasons for the time limit – nothing earlier than 1970 – have already been hinted at. As for limitation on the subject matter, it raises many questions. Is it possible to discuss human ecology but refer only to pollution, waste minimisation and recycling? Is it justifiable to write about land reform and development of the Scottish countryside, but mention only in passing matters as important as public access and reshaping of the farming industry? Should the present and future of Scotland's town and city landscapes be discussed without a review of public and private traffic management? I feel that these must indeed be possible, or any attempt at 'green' debate will fall on the need for all-inclusiveness.

One general problem, which complicates a selective approach, is posed by the nature of ecology itself. It is the interconnectedness of ecological systems that on one hand makes it difficult to focus the debate, and on the other, can make honest attempts towards green policies look like marginal, possibly insincere tinkering. This aura of insincerity hangs heavily over the assurances that 'sustainable development' is 'at the heart of government'. I have concentrated on social settings in which green thinking could have practical outcomes: create more interesting jobs and improve life-styles, as well as provide environmental benefits. Examples include the greening of tourism, planning, building and design. The greening of technology should mean less pollution, waste etc., but the outcome can also be seen in terms of interesting new products, not just cleanliness. Perhaps the most counterproductive aspect of the last few decades of environmental campaigning has been neither the shock-horror tactics nor the prescriptiveness, but the strand of profound conservatism which runs through the Green movement.

The Idea of 'the Environment'

Fashionable concepts tend to have a Cheshire cat quality: the body of thought can fade and disintegrate, leaving only the smile behind. 'The environment' is a case in point. Human rights and equity, regional identity and environmental sustainability are the grand, interrelated themes of the late 20th century. In spite of being based on simple, deeply appealing concepts, they have all turned out to be elusive in practice and bristling with internal contradictions.

Two sets of related ideas form the basis for the current concern about the environment. One set, which can be labelled 'environmentalism', deals with the fact that unless we, as a species, manage our natural assets – air, water, arable land, mineral resources and so on – in a sustainable way, we will suffer ultimately dreadful practical consequences. The other set of ideas has become embodied in the older, less alarmist 'nature conservation' movement. It focuses on people's need to commune with nature, for the welfare of their minds as well as their physical health. It usually incorporates the view that nature should mimic an ideal of unspoilt wilderness or, at least, of a pastoral landscape, rich in plants and animals. Inevitably, if surprisingly slowly, these two sets of ideas are beginning to merge to form a more or less coherent 'green' approach to everything from town planning to butterfly preservation. 'Deep ecology' is an extreme but influential version of such beliefs, perhaps most consistently expressed in the works of the Norwegian philosopher-adventurer Arne Naess.[5] The central themes in deep ecology are that good resource management is not sufficient but that 'Richness and diversity of life form are values in themselves and contribute to the flourishing of human and non-human life on Earth'; and that 'Humans have no right to reduce this richness and diversity except to satisfy vital needs.'

There is much that is purely contemporary about the notion of 'the environment' as a fragile and threatened eco-system, which includes, but must not be dominated by, human beings. Although concern about pollution, despoilment of natural resources and overpopulation are as old as large-scale industrialisation (when they replaced other environmental anxieties), and identification with the landscape is as old as mankind, the combination of anxious nature-love, scientific insight and fear of a technological Armageddon is uniquely a late-20th aggregate of ideas. This is one rationale for focusing on the present and defining it fairly narrowly.

I decided to include hardly anything written before the early 1970s and one key justification is the relative transformation of UK – and world-wide – environmental politics in the '70s. Post-war conservationist legislation, which among other things had achieved the creation of National Parks in England and Wales, though not in Scotland, was followed up by the Countryside Acts of the late '60s and a successful British participation in the European (Nature) Conservation Year 1970. This story, which reads to me like a fascinating variant on the theme of British innovative thinking, pragmatism and faltering execution, is given an authoritative treatment in T. C. Smout's edition of conference papers on *Nature, Landscape and People Since the Second World War*.[6] The contradictions of Scottish nature conservation will be explored later in this book, for instance in the contexts of the vocabulary of partnerships and sustainability (see Chapter 3), 'institutionalised nature' (see Chapter 6) and rural land-use (see Chapter 9).

It was in the '70s that the obscure Ecological Party became the Green Party, which went on to get almost a fifth of the votes in England and Wales in the local and European elections in the late '80s. The Scottish Green Party got less than 10 per cent. Although compared with anywhere in western Europe, UK voters were (and are) unusually reluctant to vote Green and the main parties are weak on environmental policies (cf. below), the '70s brought the beginnings of an eco-conscious establishment.

During the '70s and '80s, public interest in the UK focused mainly on pollution and waste. Significant legislation for England and Wales (also relevant in Scotland) included the Control of Pollution Act 1974 and Her Majesty's Inspectorate of Pollution Act 1987 and various other agencies set up to encourage resource management through conservation. Across the whole spectrum of 'green-tinged' policies, the UK was comparatively unsuccessful and the USA, Germany, The Netherlands and the Scandinavian countries led by example. By 1990, it was still perfectly legitimate for Chris Rose to write a book about *The Dirty Man of Europe – The Great British Pollution Scandal*.[7] In this survey of bad practice, Scotland figures relatively little. There are many likely reasons, but these sadly do not include that Scotland was a 'clean' country. In every case, be it high body levels of lead in Glasgow children or supine public apologists for acid rain, Scotland's showing is poor.

At the time, the serious public debate about environmental issues mostly took place outside the ken of Scottish writers and journalists, in spite of a flow of seminal works in England, including *Small is Beautiful*, *The Energy Question*, *The Common Ground*, *Troubled Waters* and *The Green Capitalists*, the last one being an early sign of the hope that big business could be made to listen and learn.[8] During this period, the main 'current-affairs' contribution from Scotland, *The Politics of the Environment*, arrived at nationalist conclusions, regardless of issue.[9] The Scottish ecologist Frank Fraser Darling's Reith lectures in 1969 were as articulate as anything written on the ecological crisis at the time, but focused almost entirely on global and UK-wide issues.[10] Almost thirty years later, the focus has started to shift in books such as *Fragile Land: Scotland's Environment*, *Scotland's Environment: the Future* and *Nature Contested*; and in studies of land reform such as *Scotland: Land and Power*.[11]

Driven by European and international directives in the wake of the UN Conference on Environment and Development (Rio, 1992), British environmental legislation became more sophisticated and inclusive.[12] In Scotland, two main legislative changes affected both nature conservation and environmental protection. One created a new nature conservation agency, Scottish Natural Heritage (SNH, 1992), which took over the responsibilities previously managed by the Scottish offshoots of the Nature Conservancy Council and the Countryside Commission. The Scottish Environment Protection Agency (SEPA, 1996) provided another organisational umbrella, in this case for pollution control authorities in Scotland so that control of air and water purity and disposal of waste could be properly co-ordinated.

It is impossible to predict whether the devolved – potentially independent – Scotland will be more or less willing to spend time, money and energy on environmental matters. For one thing, it is hard for small nations to manoeuvre in a world dominated by profoundly un-green big players. Public concern can be vigorous, for instance on issues related to recreation or human health, but is increasingly being channelled through non-governmental organisations (NGOs). Many are single-issue based and this can generate confusion and conflict as well as dynamism. It may also be true that concern about nature is a cyclical phenomenon. 'International crises' writing started with a few books such as *Silent Spring* and continued with *The Population Bomb*, *Blueprint for Survival* and *The Limits to Growth*.[13] But the sense of across-the-board urgency has faded since then, although isolated problems (e.g. climate change) receive much attention. 'The old environmentalism [has] hit its limits' as Tom Athanasiou puts it in *Slow Reckoning – The Ecology of a Divided Planet*.[14] Athanasiou's book is intended as an aggressive reminder that the First World should not risk tinkering with its own environmental policies to the exclusion of the poor of the world. He is bitter about the mindset summarised in the quote 'The world might be going to hell, but there's a lot of really neat stuff you can buy!'[15] Many other politically committed environmentalists agree that the last great international manifestations of concern, the UN-sponsored Brundtland Commission (1984–7), the Rio Earth Summit and the post-Rio wave of activity by signatories of its *Agenda 21* may have left everyone feeling rather too good about progress.[16] Andrew Rowell goes further in his book *Green Backlash* and speaks about what he calls an industrial-scale 'green-wash' (as in 'whitewash').[17] He convincingly describes not only the green-washing, but also what he calls 'a global subversion of the environment movement'.

It is impossible for any interested observer not to be struck by the amount of green window-dressing found everywhere. Still, there is a more optimistic version and the best possible scenario is at least possible. If the *Agenda 21* model could only work, the global action plan of the 'old environmentalists' might be replaced with intense local activity, involving the participation of millions of informed individuals, organised in thousands of different ways. There are places and nations which make this seem possible, and it is worth trying to understand why Scotland is not yet one of them. Comparisons with 'abroad' can be odious and irritating, but in telling the Scottish story a wider context might be useful at times. Scotland's conditions are special to it, but not radically different from a group of northern European states, which includes the Scandinavian countries, The Netherlands and the German Länder.

Social Representations of the Environment

The first First Minister for Scotland has provided me with one example of what this book is about. In Donald Dewar's introductory remarks to the handsomely produced *Corporate Scotland 1998/99*,[18] he explains that Scotland is ready to 'meet the challenges of the Millennium' and lists many good things to prove his point: the new Parliament, the export potential of old and new Scottish industries, inward investment, tourism. These assurances are illustrated by a photograph showing a beautiful Scottish view, dominated by the sky at sunset and the jutting black bulk of Arthur's Seat. The silhouetted roofs and towers of central Edinburgh run along the lower edge like a delicate fret-work. But none of this is referred to in the text. The sky above Scotland, its cities and hills have been used – yet again – as eye-catching patriotic vignettes. Or, as Donald Dewar put it in his only reference to matters environmental: 'Scotland continues to be an attractive place to invest in and do business.'

Men and women of action can afford to forget about syntax, but the consequences of reducing Scotland's landscape to a pleasing backdrop will surely soon become too costly. Perhaps the most effective counterbalance lies not in one particular set of actions and attitudes, but in the fact that people everywhere are beginning to form new and different 'social representations' of nature. The idea that members of a society collectively embrace certain ideas, images and practices – social representations – and that these in turn interact with the way that society works, comes from social psychology.[19] We are all becoming more aware of the malleability, for better or for worse, of our environment. The older idea of nature as stable and external has changed and one is now to view it as a dynamic and inter-active system that no longer allows passivity. Or, at least, this is the optimistic theory – in fact, radical change needs more than insight. It seems that not even international-scale catastrophes, such as Chernobyl or Indonesian rainforest fires, have more than marginal social effects, even in the affected countries.

In the 1960s social scientists began studying relationships between people and their 'non-social' environments in earnest. There are still few studies of attitude shifts in response to well-documented environmental threats or disasters among the many books that deal with concepts of nature, and very little about work based in Scotland.[20] An interesting doctoral project, which examined the Shetlanders' response to the *Braer* oil spill in 1993, emphasised how people's attitudes to the environment are linked to their sense of belonging to a social category.[21] The local reaction seemed to correlate (no statistical evaluation was attempted) strongly with, for instance, 'original inhabitant' or 'incomer' status, and with the main source of family income. The attitude to the spill often formed part of a greater complex of views on, for instance, the management of the oil income by the Shetland Islands Council. Attitudes to traditional aspects of the Shetland environment, for instance use of land for grazing, the lack of trees or fishing practices were less predictable.

For everybody, it is a hard act to balance between deriving a livelihood from a particular place and wanting to protect it simply because it is *your* place. Such conflicts breed inertia, bunker-mentality and the NIMBY-syndrome (Not In My Back Yard). These are complex social issues and it would be interesting to see more deliberate use of the skills of social scientists in examining them. The slow mills of academic publishing tend to negate the potential effect of the work; the analyses of current cases such as the 'super-quarry' on Harris, and the restrictions of deep-sea fishing are apparently years away from publication. As it is, the relationship between the people of Scotland and their environment cannot be approached by reading up research results. Historians are building a solid platform of observation of the past, but there is little evidence about current Scottish views on nature, except for what can be fished up in cultural cross-currents or observed to happen in Scotland's towns and countryside.

A Nation Built on Paradoxes and the Politics of the Environment

Over the last decade or so, the neglect of Scotland's environment has looked like being replaced by a more benign – and realistic – approach. The most striking changes have been in the 'greening' of the language and outlook of central and local bureaucracies, and in the growth of environmental studies in Scottish universities. But on the whole, nature in Scotland remains the subject of cultural paradoxes: it is beautiful but static, talking about it is either boring or sentimental (or both) and may be good for health but is usually bad for progress. The landscape is an essential part of national identity but still marketed as a commodity, and its existence is a constant with no dynamism of its own. Most Scots express admiration and even love for their 'native stones', but it is rare for this affection to be translated into active concern.[22] It is a journalistic commonplace to speak of 'Scottishness' in terms of paradoxes. The Caledonian identity itself is a contradictory abstraction and paradox is 'the key to the Scottish genius', which seems to be turning inconsistency into a virtue.[23]

So why should one of Europe's most homogeneous and tradition-bound populations, living in a very beautiful country, fail to look after its environmental assets? It could simply be that for many generations now, since the Industrial Revolution took hold in the 1820s and '30s, most of Scotland's people (some 80 per cent) have been living in cities. Even parks, let alone tracts of real countryside, must seem alien places to most urban Scots, who are mainly concentrated in the Central Belt. For long now, they have tolerated a whole range of disturbing social problems, from a quasi-feudal structure of land ownership to pollution – be it open-cast mining or dirty beaches – and exploitation of the countryside, from grazing to tourism. Could it be alienation from the land that has caused a nation that prides itself on science-based inventiveness to have ignored the potential of 'green technology' and at the same time caused a people that loves literature to have so few writers with nature as a main subject? It cannot be the whole answer, and I turned to the political scene for more clues.

Because modern democracy tends to make politicians reflect rather than lead opinion, public political discourse (as opposed to the backstairs variety) is normally the sum of popular attitudes and cannot be looked to as an engine of cultural change. Manipulation of economic circumstances is the one important activity in which politicians consistently take the lead. Even here, their eyes are usually firmly fixed on how matters can be presented to coincide with the conventional wisdom of the electorate. Unsurprisingly, 'partnership' and 'consultation' have become the key responses – often avoidance responses – in contemporary debate.

If all this is true, the weak political base of Scottish environmentalism is yet another consequence of its low status in the nation's cultural awareness, rather than a characteristic of Scottish politicians in particular. Pro-environment interventions may well improve as the devolved Parliament gets established, but it is much less likely without an impetus from the public. Even a cursory examination of the policy statements suggests that all the parties are feeling uncertain how to formulate an environmental agenda. Even the Scottish Green Party seemed a bit less precise than expected.

Behaving like any interested citizen might, I relied on the most easily available sources to review green policies. I have looked at the party web-sites and read what leading political figures said in the run-up to the 1999 Scottish parliamentary elections, mainly relying on a competent compilation called *Essential Scotland*, published in November 1998.[24] In it are two-page statements in small print by Donald Dewar, Alex Salmond, Malcolm Rifkind (David McLetchie had not emerged at the time), Jim Wallace, Graham Leicester (Director of the Scottish Council Foundation) and John Lloyd (a political journalist). Finally, to fill some of the many remaining gaps, the party representatives manning the telephones in the Scottish head offices were asked a brief set of standard questions.[25]

The statements in *Essential Scotland* completely ignored the environment. The E-word itself was used only once, in the context of 'business environment'. Even businesses depending on intelligent management of nature such as farming and fishing were given only passing reassurances.

The web-sites were almost as terse about environmental policies, although the Conservative aim to capture the countryside vote had actually resulted into a quite impressive array of rural policies. Most of the parties (but not Labour) tried to reassure the electorate about their own favourite pollution controls and to say something about 'integrated traffic management', but not much more. An interesting oddity was that almost everyone (except the Greens and the Socialists) felt that lifting the ban on beef-on-the-bone – the ban was introduced in the wake of the BSE (bovine spongiform encephalopathy) disaster – would be an environmental policy vote-catcher. This has got to be a particularly craven example of politicians following rather than leading.

The results of the attempted telephone interviews were negative. The party representatives did not like to commit themselves and offered to send printed material instead. The answers focused on claims that the parties would get to grips with pollution and traffic management. On the contentious issues of GMOs (genetically modified organisms) being grown in Scotland (I tended to suggest this and wasteful energy use as 'Scottish problems'), responses were cautious; for instance, most parties wanted a 'moratorium' on GMOs, i.e. 'wait and see'.

When the Scottish Labour–Liberal coalition had formed the government and been 'outed' with respect to the policies they would prioritise, three environmental Bills were proposed, dealing with National Parks, land reform and traffic congestion charging. The National Park and the land reform proposals will be looked at more closely later (see Chapter 6). The traffic congestion charges were likely to cause trouble all round: while not radical enough for some, they were also criticised as 'unfriendly to motorists'. The idea looked like a hostage to fortune and has been dropped since.

At the time of writing other changes have emerged. The National Park idea has been turned into real legislation, but it looks as if nature may have to give way to tourism (see Chapters 5 and 6). The land reform proposals are in the wings, waiting and meanwhile being targeted with cogent criticism from the gallery. Rural development has moved into the forefront of the debate as countryside crises – livestock disease, collapse of the dairy industry, the shaming of fish farming, more rows about GMO planting – have become a national concern. Scottish politicians are as aware as any of the risks entailed in setting out policies without immediate economic benefits and see little point in pushing costly (at least in the short-term) eco-programmes in a country with little interest and less education in these matters. Still, 'the environment' and 'sustainable development' have become seen as universally agreeable as motherhood and apple-pie.

Nature and Identity

The closeness past generations felt with the reality of nature has become a rarity and it seems to be history rather than the landscape that cements identity in Scotland, but those iconic landscapes must surely still be elements in the self-image of the Scottish people. It is related 'social representations of nature' that I shall try to trace in the rest of the book. One Scotsman's ominous version is quoted here:

> It was a day peculiar to this piece of the planet,
> When larks rose on long strings of singing
> And the air shifted with the shimmer of actual angels.
> Greenness entered the body . . .
> And what did she have to say for it?
> Her brow grew bleak, her ancestors raged in their graves
> As she spoke with their ancient misery:
> 'We'll pay for it, we'll pay for it, we'll pay for it!'[26]

1. *Evening sun shines through pine trees on the banks of Loch Morlich in the Cairngorm mountains.* © Colin McPherson

Notes

1. *Lanark* by Alasdair Gray, 1982, pp. 50–1.
2. There is a definition of 'quango' in Chapter 3. The irritation with these agencies of the Scottish Executive has led to calls for a 'bonfire of the quangos', e.g. by the Shadow Scottish Secretary George Robertson (1996) and later by the SNP leader Alex Salmond.
3. The interview with T. C. Smout is summarised in Chapter 4. A major work in environmental history is *Scotland since Prehistory. Natural Change and Human Impact* by T. C. Smout, 1993.
4. *Scotland – the Brand. The Making of Scottish Heritage* by David McCrone et al., 1995.
5. One key work is *Ecology, Community and Lifestyle: Outline of an Ecosophy* by Arne Naess, 1990.
6. *Nature, Landscape and People since the Second World War* by T. C. Smout (ed.), 2001.
7. *The Dirty Man of Europe – The Great British Pollution Scandal* by Chris Rose, 1990, e.g. pp. 96–7 and 127.
8. *Small is Beautiful* by Ernest Schumacher, 1974; *The Energy Question* by Gerald Foley, 1976; *The Common Ground* by Richard Mabey, 1980; *Troubled Waters. Rivers, Politics and Pollution* by D. Kinnersley, 1988; *The Green Capitalists. Industry's Search for Environmental Excellence* by John Elkington, 1987.
9. *The Politics of the Environment. A Guide to Scottish Thought and Action* by Malcolm Slesser, 1972.
10. *Wilderness and Plenty. Reith Lectures 1969* by Frank Fraser Darling, 1970.
11. *Fragile Land: Scotland's Environment* by Auslan Cramb, 1998; *Scotland's Environment: the Future* by George Holmes and Roger Crofts (eds.), 2000; *Nature Contested* by T. C. Smout, 2000; *Scotland: Land and Power* by Andy Wightman, 1999.
12. *Environment and the Law. An Introduction for Environmental Scientists and Lawyers* by John McEldowney and Sharon McEldowney, 1996, e.g. p. 8 and pp. 36–53 (European and Environmental Law).
13. *Silent Spring* by Rachel Carson, 1964; *The Population Bomb* by Paul Ehrlich, 1968; 'Blueprint for Survival', *The Ecologist*, 1972 2 (1): pp. 1–43; *The Limits to Growth* by D. H. Meadows et al., 1972. Note the follow-up, *Beyond the Limits to Growth. Global Collapse or a Sustainable Future* by D. H. Meadows et al., 1992.
14. *Slow Reckoning – The Ecology of a Divided Planet* by Tom Athanasiou, 1997, e.g. pp. 7–8 and 38.
15. From a review in the magazine *Wired*, 1993.
16. *Agenda 21* is a comprehensive plan of action, specifying local and national actions, as well as global (e.g. managed by UN agencies) in every area in which there are human impacts on the environment. *Agenda 21*, the *Declaration on Environment and Development*, and the *Statement of*

Principles for the Sustainable Management of Forests were adopted by more than 178 governments at the United Nations Conference on Environment and Development (UNCED) held in Rio de Janeiro, Brazil, 3–14 June 1992. The Commission on Sustainable Development (CSD) was set up (December 1992) to ensure effective follow-up of UNCED. CSD is to monitor and report on implementation at all levels. The United Nations General Assembly reviewed progress in 1997 and the 55th General Assembly session decided in December 2000 that the CSD would serve as the central organising body for the 2002 World Summit on Sustainable Development, which will be held in Johannesburg, South Africa. An electronic copy is available on www.igc.apc.org/habitat/agenda21/index and the annotated text in *Agenda 21: Earth's Action Plan* by Nicholas A. Robinson (ed.), 1993.

17. *Green Backlash. Global Subversion of the Environment Movement* by Andrew Rowell, 1996, e.g. p. 320ff.
18. 'Scotland Ready to Meet Challenges of the Millennium' by Donald Dewar, in *Corporate Scotland 1998/99*, by Bill Magee (ed.), 1998, pp. 2–3.
19. 'The phenomenon of social representations' by Serge Moscovici, in *Social Representations* by R. M. Farr and S. Moscovici (eds.), 1984; *Environment and Behaviour. A Dynamic Perspective* by Charles J. Holahan, 1978; *Handbook of Environmental Psychology* by D. Stokols and Irwin Altman (eds.), 1987.
20. *The Idea of Wilderness. From Prehistory to the Age of Ecology* by Max Oelschlaeger, 1991; *Redefining Nature. Ecology, Culture and Domestication* by Roy Ellen and Katsuyoshi Fukui (eds.), 1996; *Uncommon Ground. Rethinking the Human Place in Nature* by William Cronon (ed.), 1996.
21. *Social Representations of Nature: The Case of the Braer Oil Spill in Shetland* by Marie-Claude Gervais, doctoral thesis at London School of Economics, 1997.
22. *Native Stones. A Book about Climbing* by David Craig, 1987, e.g. p. 6.
23. 'The Scottish Genius' by Alan Massie, in *Essential Scotland. Seventy Perspectives on the New Scotland from Leaders in their Fields* by Maurice Fraser (ed.), 1998, pp. 48–9.
24. Ibid, pp. 7–21.
25. Do you feel your web-site adequately represents the policies of your party? Are the environmental policy statements complete? What are your party's top priorities in the environmental policy? What does your party see as the main Scottish environmental problem?
26. From 'Weathering' by Alastair Reid 1978, quoted in *Scotland – An Anthology* by Douglas Dunn, 1991, p. 28.

2

Literature and Perceptions of National Identity

Landscapes of the Mind

MOST OF US have a landscape that we think of as 'ours', kept as a mental touchstone for home. It can be reconstructed in many moods, from hostile recall to longing poetry. Mostly, it lives in the mind as a series of images remembered with low-key affection, as in this quote from Candia McWilliam's *A Case of Knives*:[1]

> Past the green flocking hills always between the walls of dry stones which are clapped together at the top like hands, past big harled houses with too little window and slipping turrets at their shoulder, and not with flowers but with green vegetables at the south walls. If any of it was to disappear, I would still see it, for I see it, of course, with sentimental eyes. What would another make of it? I am partial.

But such 'partiality' can be potent enough to create an anchoring point for a sense of identity with a place or a country. For some people and in some settings, identification progresses to another, more general stage dominated by a desire to understand how the environment functions, what it is and how the human and non-human interact.

It is impossible to escape 'the identity question' these days and later in this chapter I will come back to current thinking about the elements of Scottish national identity. If the landscape matters, then rootedness in a defined place should be a powerful motive for writing about places, what they look like and how they function. It has been important for the argument of this book to try to trace representations of place and environment in new Scottish writing. This turned out to be quite difficult.

Generally, in the phenomenon that became known as the 'New Scottish Writing', both the urban and rural landscapes seemed distant and elusive. Given my myopic aim of finding an environmental dimension, the selection I have made is necessarily idiosyncratic, but I have tried to cover as wide a range as possible. Scots living away from Scotland, like William Boyd and Andrew O'Hagan, who write about Scotland have been included. On the other hand, I have not tried to investigate Scottish-based writers using Gaelic or Scots or Doric, since their exclusion of most English-speaking readers suggests that these writers are pursuing a very special local discourse. After much agonising, I decided to exclude poetry for the depressingly simple reason that it is not much read and my main concern is looking for ideas of the Scottish environment in society at large.

Environment and Imagination

A general idea of what I have been looking for was set out in an earlier effort of mine to explore contemporary Swedish literature for representations of the environment: '[writing] in which the environment is integral to what happens to people'.[2] This is writing not exclusively focused on human relationships, but also on interactions between people, places, objects and non-human creatures. The world around you, if you care about it, is a rich source of food for the mind. Imaginative understanding of the environment is about more than descriptive prose passages.

I felt that in Alasdair Gray, perhaps especially in *Lanark*, I had found an absorbing, articulate and ultimately baffling variant of the Scottish refusal to engage with the 'world outside'.[3] Cairns Craig, in his wide-ranging study of the modern Scottish novel, refers to the way in which Gray's characters, real and virtual, 'journey to Never land', and goes on to say: 'The reason that imagination kills the world is because under modern conditions – or under Scottish conditions – the imagination is simply an escape route from an unacceptable reality'.[4]

In *Lanark*, Duncan Thaw feels that Glasgow is a city 'where nobody imagines living'. It sounds like a despairing counterpoint to the idea that 'Imagination is our link to reality'.[5] I believe literary fiction is both a reflection of society and a cause of social and political effects. It follows that the way Scotland's public affairs are managed must be related to the *Zeitgeist* generated by the people, including the writers. The pendulum may be swinging away from human interactions to places and processes, but it seems that most of the credit must go to a handful of journalists and academics. Still, why should writers be engaged in such abstractions? Surely A. L. Kennedy is right in saying, as she did in an email exchange with me (see later): 'Obviously, most literary fiction will be homocentric, it is difficult to make non-human protagonists and – if you are dealing with fiction and poetry there is a limit to what the form will bear in the way of lecturing and polemic'?

Yes and no – in just one sentence this clever writer manages to pick up the wrong end of the stick, not once but three times. It is worth dealing with her points carefully, particularly since the general feeling that 'I don't understand what you're looking for anyway' seems implicit in some of her responses.

First, the 'homocentricity' of literary fiction is indisputable, just as the 'chemocentricity' of chemistry is not an arguable quality. It is surely equally obvious that prose, whether written by literati or chemists or anybody else, can relate to a multiplicity of topics. Representing environments in literature does not mean somehow excluding people.

Second, the idea that anyone should positively want 'non-human protagonists' is more amusing than controversial. Granted, there are Swift's Houyhnhnms, Orwell's *Animal Farm* creatures and a few other examples of notable literary animals – usually only too human – but generally it seems a daft idea, and very difficult to sustain. 'If-these-stones-could-speak'-devices are even more awkward to handle outside the creative writing class.

Third, there is the concern that including the environment as a political or moral issue would mean weighing down 'the form' with non-literary arguments. Surely the notion that social messages distort prose fiction is out of order: what about Swift or Orwell for a start? Nor is it possible to argue that modern English-speaking writing has completely abandoned such concerns in favour of 'homocentricity' – there are people like A. S. Byatt on biology, Iris Murdoch on environments, Colin Thubron on places and many more.

The Scottish Countryside in Literature

In his anthology of prose and poetry about Scotland Douglas Dunn says that he has taken as his motto the maxim that 'All history is contemporary history'.[6] Although his selection starts with presumably ancient pieces by Anon, readers must assume that this most well-read of Scotsmen has picked literary references that still resonate with Scottish people. Among the eleven chapters, from 'A Strong Scots Accent of the Mind' (national character) to 'Diaspora – Scots Abroad', 'the countryside' is represented in a chapter with the slightly weary title 'Nothing but Heather'. Of course nature turns up elsewhere in the anthology, but mostly in the familiar representation – the beautiful, inspirational, nostalgia-inducing stage-set behind the action. This is actually true for most of the 'Heather'-pieces as well, which tend to deal with the emblematic, like significant plants (e.g. thistles), place-names and historical landscape scenes.

It turned out to be hard finding contemporary writers interested in the meanings traceable in the Scottish rural landscape. Robin Jenkins in *Poor Angus* is one of the exceptions.[7] This witty account of people hovering between faith and uncertain practicality has a potent 'outside', the unspoilt landscape on an island somewhere off the Argyll coast. The landscape is essentially unknowable and possibly magic but also full of smelly rustic reality. In the story-telling, the people, the landscape and the 'non-human protagonists' (notably the cows and their attentive bull) all interrelate in various ambivalent ways.

In a completely different mood and style, the people in Andrew Greig's *When They Lay Bare* are part of their Border landscape, sometimes to an alarming extent.[8] It is a slowly unfolding melodrama, with characters rooted in the hills and streams and dark woods in a way we now tend to think belonged to a distant past, when belief in sprites and goblins animated the world outside. For Greig's characters the Border country, where communities have a long tradition of brutal revenge, has created a landscape of the mind. Wild but deceptively beautiful, it is a theatre of fear more than of trust and belonging. Interestingly, Cairns Craig starts his analysis of the modern Scottish novel with the theme of fearfulness and argues that the fear is driven as much by unforgiving community norms as by belief in a vengeful deity: '"God-fearing"; there is no more powerful term of approbation in the language of Scottish presbyterianism.'[9]

But mythologising nature can be done without relying on fear as the animating force: curiosity, delight and wonder, even helpless love are possibilities that are evoked in Niall Duthie's eclectic *Lobster Moth*.[10] The Edwardian biologist-soldier around whose reminiscences the book is centred has a fascination for moths that energises a whole mental world of learning, personal memories and fantasy. Sometimes all three come together, for instance when, at the end of a sequence of speculation on relationships, manners and mores (can a scientist be a gentleman?) and biological interest, our hero eats an Eyed Hawkmoth larva. A sensual imagination is one of the critical ingredients in this story, but nature is a presence in its world and so takes part in inevitable confrontations – but maybe that sense of inevitability has almost vanished in the here and now. The introductory quote from *The Butterflies and Moths of Europe* begins blithely: 'Almost every boy has hunted butterflies through the woods and fields, or has reared silkworms or other moths [. . .]', but that was written in 1903.

If personal experience of the landscape really is vital for relating to it, then one reason that it does not figure in Scottish novels is that the writers – like most of the population – are city-people. Andrew Greig has written two mountaineering books.[11] Robin Jenkins, whose *Poor Angus*[12] must have some relationship to his current life in Argyll, set his earlier novels primarily in Glasgow, a fact presumably due to his growing up in Cambuslang. Duncan McLean's quietly funny book *Blackden* about a young man coping with life in Aberdeenshire has been written with the special insight granted to a young man from Aberdeenshire.[13]

Sometimes the imprint of writer's self is more explicit. From the perspective of the early 21st century, the lives reflected in writing about rural landscapes are those of older writers, mostly outside my timeframe, including those who were part in the 'Scottish Renaissance' in the 1930s.[14] Neil Gunn, born in Caithness in 1891, was one of them: the son of a poor crofting fisherman, his spiritual home was the north coast of Scotland. A generation or so later, Gavin Maxwell grew up in an aristocratic family seat, but he too learnt to love nature, from the lonely Lowland moorlands of his youth, to Sandaig, the place on the north-west coast he made his own. Observed with an intensity more fantastic than scientific and the subject of his finest writing, this refuge was ruined as much by the writer himself as by

2. *Deerstalkers on the privately owned Auchlyne Estate near Crianlarich, Stirlingshire.* © Colin McPherson

interfering humanity at large. It was the environmentalist's nightmare scenario – we destroy all that we love best – and it turned out that nothing, not even a well-meaning wildlife charity, could save the sanctuary he had tried to create.

There are other examples of writing with a factual base that has a real understanding of, and engagement in, the meaning of landscapes. A generation or so after Maxwell, the poet Alasdair Maclean returned to his childhood on the Scottish coast in an autobiographical book set in Ardnamurchan. Maclean's first memories are those of a poor crofter's son, and the book is based on two interlocking diaries, his father's and his own, kept when Maclean as a grown-up went back 'home' to write during a few summer months. His sense of loss of his roots is acute and loss of the landscape is part of it. Although he blames locals and incomers alike for insensitivity and exploitation of what was once there – the machair, the woodland – he tries to be reasonable about it. Too reasonable, some would say, as he even invokes the odd but popular argument that: 'after all, nature itself is a great destroyer of environments.'[15]

Books like Ian Mitchell's record of sailing round the Western Isles are difficult to fit in anywhere: intensely personal, full of love of the Isles and concern about its people, it is still unquestionably 'a polemic' and its topics will be discussed in Chapter 3.[16] Like many others who are still prepared to integrate their experience of place with their personal message, Mitchell mainly understands the landscape as a visual spectacle. His real but never confronted problem is merging his laissez-faire, the-locals-know-best arguments with his experiences of contentedly narrated and beautifully photographed landscape: 'The mighty Blaven: sunset over Skye from Loch Slapin on the evening I roasted the Rum venison on the beach.' But then there are the people: 'This book is more about people than places'. This is one way of formulating the problem.

The Scottish City in Literature

So, the city must be where the literary action is. Not only are the homes of most Scottish citizens in urban areas, but urban playgrounds are also the breeding grounds for the new Scottish culture. How do they appear in fiction: Glasgow and Edinburgh, Dundee, Stirling, Aberdeen, Inverness and all the rest? Rather indistinctly, is the answer, even though they are full of interesting people anxious to write about them. The descriptive pieces are often just postcard views or lists of street-names or other minimalist devices to establish a setting.

Glasgow has been the subject of some of the most evocative recent writing; the city seems to create a warmth of feeling that transcends all the usual requirements on 'home' being pretty or comfortable. This passage from *Its Colours They Are Fine* by Alan Spence is a case in point, picked from many in this wonderful book:[17]

> I sit at the window looking down over Hill Street, out across the city. Glasgow.
>
> Directly opposite, across the road, a row of grey tenements, crazy-tilted chimney pots, a tangle of television aerials . . . Through a gap I can see as far south as the Renfrewshire hills, fading today into grey rainmist. (Out of that mist perhaps came the first straggling settlers to this valley, this green place.) Jutting cranes of the shipyards. Monolithic towerblocks.

This is archetypal Glasgow, with an affectionate insistence on 'what it's really like'. Alan Spence kept writing about Glasgow and it is interesting to note that he seems to become less concerned with 'place' over the years. Is it a fashion thing? Or has the city itself passed some kind of line of acceptability, beyond which its environments no longer offer much for the imagination to feed on, as Alasdair Gray seems to feel? Spence, in his recent novel *Way to Go* – a biting account of death (of the soul?) in the late 20th century – comes no closer to the city than snapshots of pub interiors, rusting car-wrecks and the shiny aluminium of the Conference Centre.[18]

Even in these post-industrial days, when the old jobs have gone without enough new ones to replace them, there are plenty of Glasgow books about 'the working classes'. Some of this

3. *The cranes of the Kvaerner shipyard at Govan on the Clyde in Glasgow, seen from across the river.* © Colin McPherson

writing is rooted in the razor gang world of MacArthur and Long's novel from the 1930s, *No Mean City*. Ron McKay reworked the theme deliberately in *Mean City*, paying his respects as he writes of 1990's Glasgow drug wars and the 'resilient humour' of its people.[19] But like the journalist he is, he also sets the scene. The book is a compendium of places, often with their background history thrown in: the Gorbals then and now; police cells and the hide-outs of the homeless in all their bleak horror; ship yards, jazz clubs and bars; the Glasgow rain. *The Big Man* by William McIlvanney is another and important example of the *No Mean City*-school.[20] McIlvanney does not hesitate to present a political (or at least sociological) analysis of what has happened to post-industrial cityscapes. A place called Graithnock loses its jobs, in an only too familiar sequence, and then: 'Even physically, the town had not been so much changed as disfigured . . . Some fine old buildings . . . were demolished and where they had been rose a monumental slum they called a shopping precinct.' Later the focus shifts to the jobless, rootless people who are still around, marginally protected only by a fragile sense of community.

There are many outstanding examples of Glasgow-based writing, but the places, mostly grim and sometimes inhuman, tend to be recorded with little more than passing recognition. I feel this is true of, for instance, Jeff Torrington's Gorbals in the picaresque *Swing Hammer Swing!* and his car-factory in *The Devil's Carousel*, and of James Kelman's Glasgow, both in the philosophical *A Disaffection* and the ferocious *How Late It Was, How Late.*[21] But at least one, more recent Greater Glasgow book, Andrew O'Hagan's *Our Fathers*, is remarkable for the cityscapes and landscapes coming to life in a way that combines emotion and understanding.[22] O'Hagan is good on the saving graces of science: '"It makes sense to know how the world is alive," she said.' He is knowing about the pitfalls of politics and brilliant about buildings: 'He knew the very shadows of these walls like brothers.' These 'non-homocentric' themes are part of the human stories he is telling.

It may be the influence of the Glasgow novel or just a result of living in present-day Scotland that the Edinburgh novel too is dominated by the losers in the big city game, and that the settings can be bleak and skeletal. As Cairns Craig says with reference to (among others) Irvine Welsh's *Trainspotting*:[23] 'Charting the hell of Scotland's industrial squalor has been a fundamental requirement of the modern Scottish novel'. The popularity of *Trainspotting*, an odyssey through Edinburgh's saddest backwaters by a group of lost young people, is based on its play with language, its incongruous humour and perhaps the readers' instinctive engagement with hopeless underdogs. The urban landscape seems as intractable and as pointless as the human beings drifting through it.[24]

This harsh Edinburgh is also present in the best-selling detective stories by Ian Rankin. In the latest one, *Set in Darkness*, there seem to be more departures than usual from a rather minimalist descriptive routine. Normally this includes items such as slums: 'Gangs had been in and sprayed their names and their urine around the place'; street-names: 'persuade the driver to take you the last half-mile into the Garibaldi Estate';

thumb-nail sketches: 'There were doors on either side, leading either to a window view or an interior space.'[25] But this time the book opens with a tour of the site for the new Parliament building, during which Detective Inspector Rebus reflects that he ought to 'try to understand this home of his'. He quotes with sudden appreciation the first lines of Hugh MacDiarmid's famous 'But Edinburgh is a mad god's dream / fitful and dark, unseizable in Leith and wildered by the Forth.'[26]

Unlike Ian Rankin, William Boyd is a Scot of the peripatetic kind. It may be that his distance from Scotland makes for vivid remembrance. In *The New Confessions*, which is a peripatetic book, he has written an unsparing account of growing up in an Edwardian but still recognisable Edinburgh, complete with sea-side visits to Fife and boarding school in the Borders.[27] Not only in this book, but in others too – such as *Brazzaville Beach*, set in Africa – Boyd writes of places with a mixture of sensuality and detached observation of processes beyond human manipulation. Getting away from the mental *loci* of the homeland is not uncommon for Scottish writers, but it is interesting that living away from Scotland can, at least for writers like Boyd and O'Hagan, to such an extent sharpen awareness of the place.

As for Scotland's other cities, Aberdeen, Inverness and Stirling figure hardly at all in contemporary fiction, but Dundee fares better. Kate Atkinson has woven its cityscape into the very 'environmental' narrative of her playful story *Emotionally Weird*.[28] It is mainly about the middle-class (students, academics and dubiously landed gentry) and their tangled, mostly unfulfilled relationships. Dundee's topography plays a role and the River Tay is patiently present, its colour shifting between shades of grey. Another setting altogether, a wild, misty, Scottish sea-side landscape without human comforts, surrounds the one genuine rebel in the book. It seems that pottering around the city becomes the landscape equivalent of enfeebled but tolerable emotional goings-on, while out there – 'on the edge' – feelings are more elemental. A decadent commune in an almost-suburban country mansion serves as a kind of transition stage to the real wilderness. It is a clever and funny fantasy about the interrelationship between who people are, what they feel like and where they are. In Andrew Murray Scott's *Tumulus* Dundee is described with almost exasperated fondness.[29] It is also mainly set in the beer-swilling educational backwaters in the pubs round the university, but is a strange and gripping book with the city as a central theme: 'Living in Dundee is a game of the senses, requiring anaesthesia on a regular basis. Don't get me wrong. I love the dear old place and I feel I own it or it owns me, otherwise I would have described it under some pseudonym, Duntaw perhaps or Tayburgh, or like Sir Walter Scott, Fairport.'

What makes people enter into such mutually possessive relationships with places? It seems to be a cultural mesh that can hold you so close to it – be it fields and beaches or squares and back-streets – that the place becomes part of your emotions and fantasy world as well as of your rational appreciation. It is only too easy to speculate that this mesh is ripped apart by 'modern life' with its uniformity and tendency to create 'machines for living.'[30] Modernity is inimical to mysticism of the kind that refuses to believe only men and women are worth identifying with. Identification with the non-human requires a special kind of imagination, although I cannot work out if it is more related to the intense and often fearsome animation found in folk-tales or the absorbed admiration of scientific inquiry.

Although what I was looking for in writing of all kinds seemed easy to recognise when I found it, it is difficult to describe what was 'not there'. Descriptive writing is after all part of novelists' stock-in-trade, but scene-setting does not amount to 'representation of place'. Below, I have tried to show what I did not find in books that were put my way by concerned ('what *are* you looking for?') friends.

People Without Places

I had many recommendations of books to read in order to find 'Scotland, the land' in contemporary Scottish prose writing. Interestingly, most of them did not contain what I was looking for, suggesting a built-in resistance against the whole idea in my well-read advisers.

The Bridge by Iain Banks was one of the more absurd suggestions.[31] Iain Banks is a fantasist through and through and one of the least 'place-sensitive' writers in Scotland. The bridge in the book is a strange sci-fi construction and the fact that it was apparently inspired by the great railway bridge over the Firth of Forth is not important. The same sense of being disconnected from any kind of reality is present in A. L. Kennedy's extraordinary and much admired *Everything You Need*, which should not qualify for inclusion here because it is allegedly set in Wales, but is too good an example of writing about place-less people not to be mentioned.[32] *A Very Quiet Street: A Novel of Sorts* by Frank Kuppner was recommended to me as being about 'the relationship between people and the city architecture'.[33] It is mainly about research on old court proceedings and the reader is forced to listen to the narrator, an introvert who never stops talking about every aspect of the case he is working on and the details of his daily life. He is a running record-keeping system, not someone who interacts with things and people around him. *The Shoe* by Gordon Legge came with a warm recommendation of an intense engagement with a place, a small post-industrial town somewhere between Edinburgh and Glasgow.[34] Its engaging flow of words sharply illuminates the minds of a group of half lost, half grown-up men, but tells little about their urban reality.

Perhaps the most intriguing work in this group was *These Demented Lands* by Alan Warner.[35] It is about Scotland and a demented place it is, where the land and the people it attracts match each other in eccentricity. The place is invaded, named after foreign places and colonised by strange objects and people who want to use it for incongruous purposes. In the end, it turns out to be about Scotland's history, its true obsession, and not about the place.

The Historical Novel

Contemporary or not, Scottish writers love historical themes. Historical novels span the whole range from romance to oblique cultural or political commentary on the present. Admirable romances, like those produced by Dorothy Dunnett, Margaret Elphinstone or Nigel Tranter, can contain evocative descriptive writing, as in Dunnett's *King Hereafter*:

> East towards Duncansby, she stood on the broken, glittering shell-sand, white and lilac, of Sannick Bay and spoke of the ailments of geese, and of the making of tallow from sheep-fat, while the fulmars rose and fell with motionless wings in the salt airs round Duncansby Head, and she thought of another grey goose but did not speak of it.[36]

But there is one recent historical novel that at least points in a new, more inclusive direction: *The Fanatic* by James Robertson. Robertson's 17th-century Edinburgh has no easily identifiable link to the decorous contemporary city, and the land around it is alien too. The descriptive writing is unusually involving: 'The moor was a place of refuge. The boy saw that. In its endless browns and greens you could become nothing, be hidden from the eyes that sought you. You could coorie under a peat bank, in the oxter of a rock or beneath the grass overhang of a burn.' The wild countryside can provide more than refuge and solace: 'In such a place, where the land gathered like solid waves about them, a man and a woman would understand their place in the cosmos.' At another level of involvement with places, James Robertson's narrative gives the reader a sense of insight into the effects of time by having two parallel narrative lines, set in the middle of the 17th century and at the end of the 20th. In the twisting, mock-medieval wynds of Edinburgh, two different modes of existence meet: '[For some people] the Old Town was the décor for a party, a pasted-up backdrop. Carlin in the shadows looked further, deeper in. He saw . . . a gridlock of carts and horses.'[37]

The historical novels set in Scotland, from Sir Walter Scott onwards, are so many, so varied and so much loved that it is hard to escape the conclusion that it is the past rather than the present that serves as the bedrock of Scottishness, the true source of national identity.

Identifying with the Landscape

Social anthropologists and cultural geographers have created a continuing and fascinating debate about how man interacts with the landscape (environment?). After admitting that 'the distinction between landscape and environment is not easy to draw', Tim Ingold wrote in *The Perception of the Environment*:

> [Let me explain] what the landscape is not. It is not 'land', it is not 'nature' and it is not 'space'. . . Land is a kind of lowest common denominator . . . The Landscape is a world as it is known to those who dwell therein, who inhabit its places and journey along the paths connecting them.[38]

James Duncan and David Ley have compiled a collection of papers on *Place/Culture/Representation* full of observations and thought, much of it focused round a main question: 'How are landscapes and their representations manipulated by dominant groups?' An example of the kind of problems the contributors concern themselves with includes 'the landscape of leisure', commercially generated imaginary environment where '[the outlines] of leisure, entertainment and commodity have become blurred'.[39]

It is in the context of such imaginative representations that a gnomic but instinctively understandable dictum by Roland Barthes fits in: 'Myths turn history into nature'.[40] Like all such abstractions of abstractions, it can mean many things. One possible interpretation is that patriotic myths serve to project historical events onto what is seen as the national landscape and fuse the two into a whole, but this – the *Braveheart* school of landscape appreciation – seems to be relatively rare in Scottish writing. If anything, the landscape is usually seen as having caused people to love it in spite of its harshness towards human aspirations. Another way of looking at Barthes's aphorism is that it is about the persuasiveness of literature and films and other socially available fantasies, which retell rather than record events and as a result change people's ideas about what to accept as 'natural' – as 'the way things are'. This is a means of looking at literature that takes in its ability to reflect shared ideas and states of mind in a community – its sense of identity.

Perceptions of Identity

The ideas that fuelled nationalism are fading and globalisation has not only failed to engage hearts and minds, it is generally seen as a threat to practically everything except the multinationals earning very large sums of money.[41] At risk, among other things, are tradition and the sense of belonging, be it to a family or a guild or a region or a landscape. One consequence has been that investigating sources of identity has become a major preoccupation. In Britain, the English have a long and fascinating history of speculating on the reasons for being a remarkable and also – most of the time – successful nation. They continue to occupy both the commercial and moral high ground with books in the charmingly contradictory tradition of proud self-deprecation. Jeremy Paxman's humorous, wide-ranging survey of the Englishness he feels is still about, persistent in spite of everything, has become a recent best-seller.[42] The same theme is given a much less forgiving, more regretful treatment by Roger Scruton.[43] In spite of their different tone and political orientation, the journalist and the philosopher have quite coincidental views of what the English virtues and (tolerable) vices are: humour and inability to be serious; stiff upper lip and emotional constipation; honesty and a disagreeable inability to see beyond the conventional – and so rather recognisably on. A recent book by Michael Wood deals with almost exactly the same theme, but written from a historian's point of view, listing outstanding English figures in fact and not-quite fact.[44] It is an endearing book, moving easily from King Arthur (who?) to, under the irresistible heading of 'Landscapes and People', the last bowl-turner in England, and Jarrow (of the Venerable Bede and the 1929 march). The only trouble is that books such as these, with the possible exception of Paxman's, seem unable to make plausible connections with the multi-ethnic, devolved, complex 'England' of today.

Using the name 'England' instead of the more official-sounding 'United Kingdom' tends to be regarded as offensive, because it excludes the separate identities of the Scottish, the Welsh, the Northern Irish, the Cornish – but, wait, the Cornish *are* English. Or are they? Factors like geographical boundaries and racial or family characteristics often matter less than cultural constructs and images. When cultural attributes fail to preserve a group-identity things get difficult: 'It is in the attempt to recover a lost or threatened sense of relational identity in attributional terms that people come to define themselves and to be defined by others, as "indigenous".'[45]

Attempts to impose cultural parameters on a large scale succeed only when they relate to current perceptions. Nothing will ever recover past certainties. Jeremy Paxman quotes a lament by David Starkey,[46] which begins: 'England itself has ceased to be a mere country and become a place of the mind' – and so it has, at least for people like Starkey. The quote goes on: 'England indeed, has become a sort of a vile antithesis of a nation; we are similar to our neighbours [in other countries] and differ from each other.' The article was called 'The Death of England'. There are plenty of other versions of the same theme, including Scruton's *Elegy*, the gentler, more academic *The World we have Lost* by Peter Laslett and – in the relatively recent past – George Orwell's *The Lion and The Unicorn* and *The English People*.[47]

Buried inside these compilations and evocations of 'who we once were' is much wishful thinking, exposed by Julian Barnes in his witty novel about the heritage industry, *England, England*.[48] He writes about the fusion of popular icons, patriotic symbolism and cash value in a wonderfully gross experiment in 'infotainment'. It is based on market-researched English identity tags, starting with top-ten items like 'The Class System' and 'A Robin in the Snow' and so on right down to 'Not Washing/Bad Underwear' and 'The Magna Carta'. In an insightful passage on children's ability to believe and disbelieve at the same time in things like mock-ups of historical events, Barnes asks: 'Could you reinvent innocence? Or was it always constructed, grafted on to the old disbelief? Were the children's faces proof of this renewable innocence or was that just sentimentality?'

The loss of innocence and belief in English certainties has damaged Britishness. This is felt by the Scottish, who – whether they like it or not – are inextricably linked to their dominant neighbour by precisely the kind of cultural meshwork that makes for a shared sense of identity. The feeling that old England has died, and with it British national cohesion, must be a major force behind the attempts to generate new reasons for believing in Scotland. It follows from what I have tried to say above, that I think this might fail unless the Scottish can find something in their present existence – rather than their past, however colourful – that works as a unifying set of shared concepts.

Searching for a Scottish Identity

Without the past, Scottish national identity seems almost insubstantial. Suits with indistinguishable ideas are marketing a populist mixture of uneasy boasting – 'we invented the telly' – tartanry and anti-English sentiment. The Scottish identity industry is in many ways less sophisticated than the English one, and the Scottish sense of belonging seems less secure. The role of the landscape in the iconology of Scotland illustrates the divide between reality and patriotic invention. The reality is the cityscape and its uncertain and unstructured fringes of housing and industrial estates, but somehow nationhood has become identified with the wilderness visions of the calendar and the inward investment literature.

But the search for a stronger base is on, and soon the trickle of Scottish identity-writing will turn into a flood. One of the most interesting sources is likely to be a recently formed multi-disciplinary team of academics, consisting of sociologists, social psychologists, social anthropologists and political scientists recruited from several Scottish universities under the aegis of an Edinburgh-based initiative called The Governance of Scotland Forum.[49] The research programme – *Nations and Regions. Constitutional Change and Identity* – recognises the tensions that arise from 'the changing nature and dimensions of Scottishness and Englishness' but will focus on the effects of constitutional change in general and the setting up of a Scottish Parliament in particular. The anthropologists and psychologists have their own agenda, which includes charting the consciousness of contemporary Scots. They ask questions such as: 'How do Scottish individuals construe their identity?' 'What is the effect of changing identity awareness on relationships with others, notably those of different race?' 'What are the views of young people in Scotland?' If such questions are even partly answered by the five-year programme of labour, Scotland will have a remarkable mirror in which to examine its reflection.

So far, there have been few late 20th-century followers of the articulate and literate champions of Scottishness from the 1930s and few attempts to find out what it might become, in a world that at the same time seems to move towards globalisation and regional independence. There is of course *Scotland – the Brand*, which looks at the elements in Scotland's self-image,[50] an amalgam of romantic history and apparently people-free wildernesses. More recently, a group of Scottish historians got together around the theme of 'Image and Identity' in the 'variety of Scotlands' that have emerged over time since the Wars of Independence. Sadly, the identities and the images remain separate constructs for all the eleven contributors, nor do they seem to have found much continuity over time, apart from the last couple of centuries when the British imperial idea gained ground. Some of the most lively pre-Union periods were those showing the maximum amount of internal dissension, for instance the Reformation, as discussed by Michael Lynch in his paper 'A Nation Born Again?'[51] The compilation stops short at the beginning of the 20th century. Any speculations about what the past might mean for the present are discouraged; as the joint editorial preface says, with an evasiveness that feels annoyingly familiar: 'readers may like to ponder for themselves what it means to be Scottish. After all, this is your century and it is your identity!'

The 'it's-all-up-to-you-chaps' stance of the professional historians is in fact echoed elsewhere. Christopher Harvie, a multi-disciplinary team housed in one man, wonders about Scotland's present and immediate future in a recent paper characteristically called 'Industry, Identity and Chaos':

> Scotland as a progressive, experimental culture-nation marketing itself in a regionalised Europe? The 'Plain People of Scotland' led by the Cardinal, Brian the Born-Again Bus and Tabloid Jack? Scotland, the pensioner in the UK's granny flat? Or a dysfunctional, externally controlled Scotland with consoling myths, thorny ressentiments [*sic*] and underemployed young men: *Braveheart* meets *Trainspotting*?[52]

Harvie's perceptive but confusingly all-inclusive view of the most durable components of Scottishness includes respect for civic order, technical skill and ingenuity and – I think, because of his praise of Patrick Geddes ('he popularised the term ecology') and use of the sea as symbolic of Scotland ('industry, identity and chaos in itself') – a love of nature, preferably in the wild. I would like to think this was true.

There are more political reasons for embracing Harvie's anxieties about the choices open to Scotland. If only the political debate in the Scottish Parliament had been closer to that carried on among the country's intelligentsia, it might have been more fruitful and surely much more interesting to listen to. There are a few examples of writers that try to open up popular debate. One such is Jeff Fallow, whose innocently entitled *Scotland for Beginners* is a compendium of radical grass-roots nationalism illustrated with vigorously crude line drawings.[53] I suspect his view of Scotland's options is quite widely shared: 'Further Oil Prospecting, Incentives For Industrial Development, Scottish Farmers' Markets To Sell Internationally, Computer Villages, Tax Haven For Artists And Writers'. The drawing shows these tags, stuck on to rockets taking off to fly to 'The Outer Limits'. Apart from wondering at the innocent simplicity of it all, I noted that even careful scrutiny of the book revealed not a single mention of the land, the place, the environment or any related ideas.

A much more thoughtful, but still inconclusive survey of contemporary Scotland was undertaken by freelance writer Emma Wood, an articulate 'incomer' to Ross-shire from East Anglia. After ten years in Scotland, she began observing her new homeland systematically and wrote a book about the results 'incorporating a brief history of the Scots and the English'.[54] It is an interesting, personal and biased (in spite of disclaimers) view in favour of political independence from England as a necessary means to finding a true Scottish identity. Emma Wood is too intellectually honest not to include the almost pathetic-sounding caveat: 'How would we ever live with ourselves and each other if we didn't get it right this time?'

The role of politicians in opening up a future of looking after a regenerated Scotland is complicated by different perceptions of what that 'regeneration' should entail. One expert commentator, Andrew Marr, says in his insightful, if technically out-of-date (written in 1992) *The Battle for Scotland*: 'The contemporary battle for Scotland is one between opposing visions of its identity . . . : on one hand, the assertive, left-liberal, semi-independent northern European state that most Home Rulers dream of; on the other, continued inclusion in Greater Albion, the Conservative-ruled "country at ease with itself"'.[55] This was written at the time of the John Major's Conservative Party rule, but change the 'C' to lower case, and most Scots would agree that these are the main opposing, and practically irreconcilable, options.

'Most Scots', but certainly not all, because the solid realism of Marr's choice excludes nationalism, romanticism and fantasy. Home Rulers may well dream of left-liberal semi-independence, but not poets. In *Dream State*, Donny O'Rourke has collected works by twenty-five young (under forty) poets with the not quite openly admitted aim of finding an image of a true Scotland forming in the kaleidoscope of poetic imagination.[56] It is the editor's long introduction to the selection that I will focus on, since I have promised myself not even to try to step into the critical mine-field laid by the sharp-witted fellowship of poets. O'Rourke is partisan but careful, knowing but receptive; invoking Edwin Morgan, perhaps the most internationally-minded of Scottish writers, he explains that while the chosen poets are often inspired by a wish to 'acknowledge the moment' (of a possible Scottish state), the poetry is 'characterised by a vigorous pluralism'.

All this is true enough, but there is a remarkable consensus in the extracts he quotes, all the more notable in comparison with the vagueness of so many other attempts to define what modern Scotland is about. It is about being populist but abhorring commercialisation, feeling lonely in a big, bleak world and wanting company – 'Do we dare let go and lift our hands to wave to the waving citizens of all those other countries?' – and about nostalgia for another, homelier Scotland. But, for there is always a 'but', some of the most interesting writing about Scotland comes from Robert Crawford, who – although full of regrets – 'refuses nostalgia' as O'Rourke puts it. Tackling the phenomenon of Silicon Glen, Crawford says: 'among circuitboard crowsteps / To be miniaturised is not small-minded', and 'Yi're beat / By yon new Scoatlan loupin tae yir street.'

Maybe this mixture of folk defiance and pride in intellectual skills is a good meeting-place for minds preoccupied with Scottish identity. Political power as a way to cultural independence and therefore (self-)respect is a theme taken up by Robert Crawford in his discussion of the 'Scottishness of Scottish writers'.[57] He asks: 'if the respect paid to American writers, rarely discussed under the heading "English Studies" (unlike practically any other kind of writing in English) [is based on a question] of literary merit or of American economic and political power?' He surely answers his own question with 'Power, actually', in the simple act of forgetting that 'American', without further qualifications, includes a large number of different sub-groups and could even be taken to extend to Latin America. Perceptions of power do matter in cultural identifications, but when imagination comes into play, what matters most is individual evaluation.

I do think that there is such a thing as Scottishness, a kind of extended group identity, which comes from an identification with a generous but not too pluralist set of shared values and ambitions. Historical knowledge and nostalgic affection are necessary but not sufficient underpinnings of Scottishness. To become a truly supportive present-day concept I believe the intelligentsia must shift it away from the personal into structure and process. History is not only about romance but also about precedent, and the nation state is above all about shared cultural assumptions and the guidelines these assumptions dictate. And, of course, I believe that a most important part of accepting the national cause and effect will be to move Scotland's landscape, rural and urban, from its present embalmed but slowly mouldering state and make it central to both imaginative and practical aspects of the Scottish identity.

Talking to Writers

It seemed important to find out if my (tentative) conclusions are shared by others. I am no literary scholar – had I missed something? Maybe some essential foreignness had placed a filter between the Scottish texts and my understanding of them. The writing in my native country is by comparison almost obsessed with place and nature. I asked two leading writers to tell me what their views were. The 'conversations' (one was conducted by email) could hardly have turned out more differently.

Email exchange with A. L. Kennedy[58]

Alison Kennedy was helpful and precise: she was very willing to consider direct questions but had little time to spare and preferred an exchange of views by email. Although I like talks – conversations – and then to summarise what has been said, I of course accepted the written format, but tried to avoid too much by way of Questions and Answers. Instead I sent Alison Kennedy some rough and unedited quotes from my text, picked both because they were relevant to my preoccupations and because they were sometimes provocative in tone. I asked her to respond in any way she liked.

I began with a question based on the survey of books on English identity: The loss of belief in English certainties has led to a parallel loss of Britishness. This is felt by the Scottish, who – whether they like it or not – are inextricably linked to their dominant neighbour by precisely the kind of cultural meshwork that makes for a shared sense of identity. The sense that old England has died is presumably the main reason for the effort going into creating a new belief in a unifying concept of Scotland. It follows from what I have tried to say above that I think these efforts are going to fail. Do you think so too?

A. L. Kennedy replied: The above really doesn't need to be dignified by a reply – it is written from an entirely English perspective – not to say [an] extremely right wing [one] on the part of Scruton. An equivalent argument might seek to suggest that a Swede insisted on speaking Swedish only because of bigotry, or some kind of animus towards Denmark, because the two languages are really so similar that only one needs to exist.

England is experiencing an identity crisis because it has always made the assumption that Britain and England were much the same and that British and English interests were identical – this is hardly surprising, given England's history as an Empire power. This confusion has never existed in Scotland, although the usual confusions, self-destructive myths and Uncle Tom-ing one would expect from a conquered nation have all been present.

Personally, I wish England a speedy recovery – in the process of advancing an empire, it has lost a sense of its own culture, it has a far more tentative grip on English folk customs, dances and music (all of which are delightful). Interestingly, the only time in recent history when the diversity of Englishness and the very broad spectrum of other identities within Britain was discussed and celebrated was during the Second World War, when it was necessary for citizens of Britain to fight and die for somewhere they could recognise as being home and helpful for them to believe that they didn't have to deny their own identity to be included in the implied better future of Britain.

I continued with another draft passage pondering Scottish identity issues: Besides, the mixture of uneasy boasting – we invented the telly – tartanry and anti-English aggression does not provide an attractive package for people who happen to live in Scotland to add to their cultural baggage, often from many elsewheres. The Scottish Identity industry is in many ways less sophisticated than the English one, which is probably why it seems to find assimilating outsiders and incomers particularly hard. The role of the landscape in the iconology of Scotland illustrates this divide between reality and patriotic invention: the reality is the cityscape and the invention, the pastoral visions of the calendar and the inward investment literature. Given the concentration of people into cities, have the Scottish in general and writers in particular eliminated the world that is not-people altogether from their active consideration (with some notable exceptions like Andrew O'Hagan)?

A. L. Kennedy replied: I simply don't think this is true at all. Although Scotland isn't free of racism, or bigotry, statistically, it doesn't seem to suffer from it as badly as England. The actual idea of Scottish identity is now perhaps at its healthiest – quite often because of the diverse and complex portrayals of Scotland

and its people in the work of Scottish writers and artists. I think if you're not aware of this you're missing a great deal.

I can't say I'm particularly impressed by the work of Andrew O'Hagan who is primarily a journalist and tends to present a very over-simplified picture of Catholics in Glasgow – others have done this far better, although Andy tends to want people to forget that.

Literature is the word: the best way to trace the ideas of a self-defined community or group is to examine the fiction it creates, not just literary fiction, but in all media. Concepts of identity will be inherent in the stories people tell themselves. The stories will emerge not just from self-aware artists, but from politicians and journalists and scientists and civil servants – all citizens in fact, who care to speculate aloud about who they are and the nature of the place where they live.

I then asked whether she agreed or disagreed that contemporary writing in Scotland excludes stories about that which is not human.

A. L. Kennedy replied: No, I don't at all agree – I think this again reflects a lack of knowledge of Scottish writing – there are many urban writers, but they don't deal exclusively with urban experience and there are also a very great many writers dealing with non-urban experience. Obviously, most literary fiction will be homocentric, it is difficult to make non-human protagonists and – if you're dealing with fiction and poetry there is a limit to what the form will bear in the way of lecturing and polemic.

Finally, I quoted a passage about the possible place of the landscape in (national) literature: The broad definition of what I was looking for in the shifting world of real writerly imagination can be taken from an earlier effort to mine contemporary Swedish literature for social representations of nature: '[works] that where the environment is integral to the action'.[59] This definition is vague because it has to be. It is not just landscape descriptions that I am looking for, or existential anxieties about the fate of the planet, or socio-political realism. 'Integral' means that the writer is not exclusively focused on human behaviour, but has, at some level, an engagement in interactions between people and their environment, urban or rural or anywhere in between. What do you think? Or is it that 'Imagination is our link to reality'?

A. L. Kennedy replied: I don't think any good writing lacks this engagement and I don't think the good writing in Scotland lacks it, or ever has lacked it, particularly in the area of verse, particularly in the tradition of Gaelic writing, but also in English language fiction from the time of Hogg onwards. If you want a really authoritative opinion on these matters, Robert Crawford is a fine writer and also an exceptionally well-informed academic.

Interview with James Robertson[60]

James Robertson allowed me to follow my preferred pattern and turn up on his doorstep practically unprepared apart from the questions that were always at the back of my mind. The conversation was wide-ranging and, for me, fascinating.

First I asked James Robertson to comment on the small passage in his recent novel *The Fanatic* that I thought was an example of an 'environmental' way of dealing with the past: '[For some people] the Old Town was the décor for a party, a pasted-up backdrop. Carlin in the shadows looked further, deeper in. He saw . . . a gridlock of carts and horses'.[61] He said that although he could see what I was after, he did not agree that his aim was to show how the environment can be a carrier of historical awareness. Instead the city, past and present, was seen as a restrictive human construct imposed on the landscape. Symbolically, this recurred throughout the book as a contrast between the open spaces of moor, sea shore and hill-side and the murky rooms and murkier prison-cells that are part of the enforced crowding and regimenting of people.

He went on to draw my attention to a passage set in modern times: a little boy is introduced for the first time to Edinburgh, a place that is both impressively monumental and frighteningly alien. This is followed by another equally ambivalent encounter with the city when the boy has grown up and become a student at its university. It seemed that for the grown-up at least, the vibrant presence of history – 'the electricity of time' – in the city held its own enchantment, but so did the way in which he can see the city merge into the countryside and become connected with the land of Scotland.[62]

But for Robertson himself, the sense of spiritual freedom of the real countryside and the liberating significance of 'the North' (of Scotland) is crucial. Irrational or not, he said, it is a part of him. For him, the town–country duality is inescapable, and the people in towns are almost always the losers.

Is the landscape no longer part of a larger Scottish identity, at least as reflected in current Scottish writing? He reckons this may be so. He feels that this could be in part due to the urban lives of most Scots, but also that there are purely literary reasons for the intense interest in individuals' emotional relationships, and above all, a wish of writers to distance themselves from old-fashioned 'scenic' writing and worst of all, the Kailyard legacy of cosy rural fantasy.

Robertson went on to argue that among the great Scottish writers in the '30s, many took an intense interest in the countryside and in particular the Highlands. These included Gunn and Mitchison; even someone like Hugh MacDiarmid, a committed communist, could not really abide Glasgow and saw only too clearly the inhuman quality of its industrial hinterland. MacDiarmid went on to write perhaps more beautifully than anybody else about the Scottish landscape. Robertson pointed out that MacDiarmid's native place was Langholm, a Borders mill town, where as a boy he had access to rivers, hills and woodland in abundance, and described wanting to *know* about the fauna and flora. How much do childhood experiences matter? Robertson told me of an article by Philip Hobsbaum in the latest issue of the poetry magazine *Dark Horse*[63] called 'Hugh MacDiarmid and Chris Grieve', in which Hobsbaum writes: 'Chris always said Glasgow was the best town in Scotland. I pointed out that he had cursed the place up and down in several poems, but he smiled joyously, and told me I was lucky to be living among "the red cliffs of North Kelvinside".'

The identification of Scottishness with a sparsely populated Highland landscape has annoyed Lowland writers of later generations, and among the city-based, old antagonisms led to a reaction most clearly seen in intensely urban writers like James Kelman. I mentioned my feeling that James Kelman is a 'no-place' writer. Could this be a response to dull, dirty and uniform city environments? Robertson did not agree with me at all and first quoted Kelman himself writing to the effect that 'I live and work in Glasgow. This is what I write about.' He went on to say that rather than being in 'no place', Kelman, Welsh and others portray urban settings in which there is 'no place for nature'. Urban life has some compensations, though. Returning briefly to the subject of Edinburgh, he spoke of how much its connections to other parts of Scotland and to Europe have meant for the traffic of people and ideas. Outreach keeps a city vital and gives its people a sense of liberty.

We moved on to the role of science and the delight of knowing 'how things work'. Robertson thought that such knowledge was a fit subject for literature and that the drive to investigate the world in general and nature in particular was part of an 'old Scotland' that may now be seen as irrelevant.

The mental distance between investigation and animation of nature can be curiously short, and knowledge and fantasy can feed off each other. Robertson said the work of James Hogg, 'the writing Ettrick shepherd' was a case in point. Often, people take nature far too seriously. In Scandinavian writing the virtual inhabitants of forests and moors could be comical, naughty or silly, as for instance in *Peer Gynt*. He went on to tell me about 'brownies'. They are mythical, troll-like creatures who inhabit people's houses and do a lot of household tasks overnight. You might try to leave some food for them, but they could be surly and bad-tempered if you tried to thank them, up sticks and go off to someone else's house. James Hogg wrote a novel called *The Brownie of Bodsbeck* set in Covenanting times, which features brownies and other might-be beings, described with humour and humanity.[64]

The two interviews illustrate the different sources of a sense of belonging in Scotland: A. L. Kennedy draws almost exclusively on social and literary ones. Nature, animated or analysed, appears to be of little concern to her, and this may be how most Scottish people approach 'the environment'. This does not mean that the landscape has been eliminated from Scottish consciousness, as James Robertson showed.

Of course, it would be silly to expect uniformity of views between two individuals just because they are both serious writers at a time when their country is going through a period of rapid change. Cultural contrasts may be sharper in Scotland, where the 'key to the Scottish genius' has been defined as paradox or 'Caledonian antisyzygy', as Norman Wilson called it when trying to define the contemporary Scottish writer in his introduction to *Scottish Writing and Writers*.[65] Scottish writing is indeed characterised by 'remarkable diversity of talent and purpose', but Douglas Dunn formulated the idea more precisely in his carefully non-partisan analysis of *Divergent Scottishness*: 'Deracination and spiritual homelessness, as well as the opinions and perspectives of a minority within a broadly-defined society, suggest themselves as authentic subjects for a certain kind of Scottish writer.[66]

Listening carefully to the voice of Scotland speaking through its writers has made it clear that the landscape – the place, the environment – is not a serious concern for the majority.

Notes

1. *A Case of Knives* by Candia McWilliam, 1988, p. 248.
2. 'Introduction', in *The Environment in Contemporary Swedish Writing* by Anna Paterson (ed.), 1997, p. 2.
3. *Lanark* by Alasdair Gray, 1982.
4. *The Modern Scottish Novel: Narrative and the National Imagination* by Cairns Craig, 1999, p. 231 (on Gray's writing).
5. 'A Defence of Poetry' by Percy Bysshe Shelley, 1821, in *The Percy Reprints, no. 3. Peacock's Four Ages of Poetry, Shelley's Defence of Poetry, Browning's Essay on Shelley* by H. F. B. Brett-Smith (ed.), 1929, pp. 23–59.
6. *Scotland – An Anthology* by Douglas Dunn, 1991, p. 1. The quote is from *History, as the Story of Liberty* (orig. *La storia*) by Benedetto Croce, 1941.
7. *Poor Angus* by Robin Jenkins, 2000.
8. *When They Lay Bare* by Andrew Greig, 1999.
9. *The Modern Scottish Novel: Narrative and the National Imagination* by Cairns Craig, 1999, p. 376 (on fear).
10. *Lobster Moth* by Niall Duthie, 1999, e.g. pp. 120–3.

11. *Summit Fever* (1985) and *Kingdoms of Experience* (1986) both by Andrew Greig.
12. *Poor Angus* by Robin Jenkins, 2000.
13. *Blackden* by Duncan McLean, 1994.
14. *Literature and Oatmeal. What Literature has meant to Scotland* by William Power in the series *The Voice of Scotland* by Hugh MacDiarmid (ed.), 1935.
15. *Night Falls on Ardnamurchan* by Alasdair Maclean, 1984, p. 88.
16. *Isles of the West. A Hebridean Voyage* by Ian Mitchell, 1999, p. 84 and caption, photograph 1.
17. *Its Colours They Are Fine* by Alan Spence, 1983, p. 198.
18. *Way to Go* by Alan Spence, 1998.
19. *Mean City* by Ron McKay, 1995. Cf. *No Mean City. A Story of the Glasgow Slums* by Alexander McArthur and H. K. Long, 1935.
20. *The Big Man* by William McIlvanney, 1985, p. 18.
21. *Swing Hammer Swing!* (1992), *The Devil's Carousel* (1996) by Jeff Torrington; *A Disaffection* (1989), *How Late It Was, How Late* (1994) by James Kelman.
22. *Our Fathers* by Andrew O'Hagan, 1999.
23. *The Modern Scottish Novel: Narrative and the National Imagination* by Cairns Craig, 1999, p. 231.
24. *Trainspotting* by Irvine Welsh, 1993.
25. *Black and Blue* by Ian Rankin, 1997, e.g. p. 17.
26. *Set in Darkness* by Ian Rankin, 2000, pp. 11–12. Quote from *The Complete Poems* by Hugh MacDiarmid, 1978.
27. *The New Confessions* by William Boyd, 1987.
28. *Emotionally Weird. A Comic Novel* by Kate Atkinson, 2000, e.g. p. 10.
29. *Tumulus* by Andrew Murray Scott, 2000, p. 39.
30. 'Das Haus ist eine Maschine zum Wohnen.' From the translation into German: Le Corbusier, *Kommende Baukunst*, Hans Hildebrandt, 1926, p. 85.
31. *The Bridge* by Iain Banks, 1986.
32. *Everything You Need* by A. L. Kennedy, 1999.
33. *A Very Quiet Street* by Frank Kuppner, 1989.
34. *The Shoe* by Gordon Legge, 1989.
35. *These Demented Lands* by Alan Warner, 1997.
36. *King Hereafter* by Dorothy Dunnett, 1982, p. 257.
37. *The Fanatic* by James Robertson, 2000, pp. 29, 153, 132.
38. 'The Temporality of the Landscape', in *The Perception of the Environment: Essays in Livelihood, Dwelling and Skill* by Tim Ingold, 2001, p. 190.
39. '"This Heaven Gives Me Migraines". The Problems and Promise of Landscapes of Leisure' by Stacy Warren, in *Place/Culture/Representation* by James Duncan and David Ley, 1993, pp. 173–86.
40. 'Myth Today' by Roland Barthes, in *Mythologies*, transl. A Lavers, 2000, p. 129.

41. *False Dawn. The Delusions of Global Capitalism* by John Gray, 1998; *Runaway World. How Globalisation is Reshaping our Lives* by Anthony Giddens, 1999; *Captive State: The Corporate Takeover of Britain* by George Monbiot, 2000.
42. *The English. A Portrait of a People* by Jeremy Paxman, 1999; David Starkey's article is referred to on p. 264.
43. *England: An Elegy* by Roger Scruton, 2000.
44. *In Search of England. Journeys into the English Past* by Michael Wood, 2000.
45. 'Ancestry, Generation, Substance, Memory, Land' in *The Perception of the Environment: Essays in Livelihood, Dwelling and Skill* by Tim Ingold, 2001, p. 151.
46. 'The Death of England' by David Starkey. *The Times*, 20 April 1996.
47. *The World we have Lost* (1965) and *The World we have Lost – Further Explored* (1983), both by Peter Laslett; *The Lion and the Unicorn* (1941) and *The English People* (1947) both by George Orwell.
48. *England, England* by Julian Barnes, 1998, pp. 83–5 (The Fifty Quintessences of Englishness) and p. 264.
49. Nations and Regions. Constitutional Change and Identity. www.ed.ac.uk/usgs/forum/Leverhulme/TOC
50. *Scotland – the Brand. The Making of Scottish Heritage* by David McCrone et al., 1995.
51. 'A Nation Born Again? Scottish identity in the Sixteenth and Seventeenth Centuries' by Michael Lynch, in *Image and Identity. The Making and Re-making of Scotland through the Ages* by Dauvit Broun et al. (eds.), 1998, pp. 82–104; the quote from the Preface comes from p. 3.
52. 'Industry, Identity and Chaos' by Christopher Harvie, in *Scottish Affairs* 32, 2000, pp. 1, 2 and 10.
53. *Scotland for Beginners* by Jeff Fallow, 1999, p. 161.
54. *Notes from the North. Incorporating a Brief History of the Scots and English* by Emma Wood, 1998, p. 151.
55. *The Battle for Scotland* by Andrew Marr, 1992, p. 232.
56. *Dream State: The New Scottish Poets* by Daniel O'Rourke, 1994, pp. xvi–xliv.
57. *Devolving English Literature* by Robert Crawford, 1992, p. 2.
58. A. L. Kennedy, a many-times prize-winning author, published three collections of stories and two novels prior to *Everything You Need* (1999) which was praised as a work of real genius.
59. Introduction, in *The Environment in Contemporary Swedish Writing* by Anna Paterson (ed.), 1997, p. 2.
60. James Robertson is the author of two collections of short stories, two of poetry and a book of *Scottish Ghost Stories* (1996). *The Fanatic* (2000) was his first novel, which was widely praised.
61. Ibid., *The Fanatic*, p. 132.
62. Ibid., *The Fanatic*, pp. 111–14.

63. *Dark Horse* 11, Spring 2001.
64. *The Brownie of Bodsbeck* by James Hogg, Douglas S. Mack (ed.), 1976.
65. *Scottish Writing and Writers. A Survey of Modern Scottish Literature* by Norman Wilson (ed.), 1977, p. 10.
66. 'Divergent Scottishness: William Boyd, Allan Massie and Ronald Frame' by Douglas Dunn, in *The Scottish Novel since the Seventies: New Visions, Old Dreams* by Gavin Wallace and Randall Stevenson (eds.), 1993, p. 156.

3

The Official Environment

Views of the Official Landscape

THE REAL LANDSCAPES of our minds are imprinted at special times of susceptibility, often but not always in childhood. We identify more with them, the more we have learnt to understand them. Other landscapes, presented for our approval by official bodies who want our support, can be persuasive fictions, like the tiny trees and people in architects' models. The difference between a novel and, say, a tourist guide or a political speech, is that the latter ostensibly tell the truth although the facts may be as selectively chosen as in a novel. There are fantasy aspects of public scenery, but when 'the landscape' leaves the aesthetic and emotional sphere of private love and national pride, the language tends to be judicious.[1]

The Cost Effective Landscape: Learning from Nature is the Scottish Office's new landscape design and management policy created to inspire innovative landscape design contributing to sustainable and best value objectives. The procedure is currently illustrated by a number of case studies and examples compiled from the trunk road estate. The Scottish landscape is extremely diverse in character. A thorough understanding of the site and its context is the means to learning from nature and maximising the landscape and ecological potential. Consideration should be given to:

how and why the landscape was formed;
how and why the landscape works;

how and why the landscape is valued, protected and its development controlled;
how and why the landscape will develop?

To help with understanding the landscape, science comes to help:[2]

> There is an increasing interest in the use of mapped data and geographic information systems (GISs) to assess visual landscape variables using reproducible methods over a wide area (Bishop and Hulse, 1994). Recent research has shown that the public's scenic preferences can be assessed objectively and quantitatively (Dearden, 1980). This research has also demonstrated that public perceptions can be related to and, in fact, predicted from environmental attributes of a more tangible nature (Buhyoff *et al.*, 1994) . . . The Belgian experience with landscape evaluation, especially in rural re-allotment projects, indicates, and international literature from a great number of disciplines or research field confirms, the necessity to speak of scenic or visual resource management (Tips, 1984)

No one could fail to recognise and approve of the serious intent. Will this systematic concern be translated into a living landscape, free to exist and sustain itself on its own terms?

Sustainability and the Official Environment

'Sustainability' has become a core concept of the last decade or so and in many ways represents a great leap forward from the managerial terminology of the '70s and '80s. That was the time when politicians and civil servants, writers and journalists – practically everybody who 'mattered' – still thought the environment was an unglamorous and impractical invention. During the last couple of decades, enough storm clouds have gathered to bring about radical changes in attitudes and vocabulary. Concepts such as ecology, environment, pollution and sustainability are now used on a previously unheard of scale. Ecology, like biodiversity and habitat, are understood to be good things. Environment is almost always linked with pollution and practically always interpreted in terms of human health. Sustainability was extended to incorporate any desired policy with an environmental impact. Now that the new century has brought the reality of climate change and epidemic sickness related to livestock, sustainability will surely be reinterpreted. Its present magic-cupboard quality is masking its efficiency in analysing real problems.

How should sustainability be defined? The simple answer is that it is about finding useful means of conserving environmental resources. It is maybe too simple. The Convention of Scottish Local Authorities (CoSLA) has been considering the matter ever since its members signed up to *Agenda 21*. Its definition is: 'Sustainability is about improving people's quality of life by integrating social, economic and environmental issues. It is about ensuring the availability of resources for current and future generations'.[3] It does go on to say that it also is about 'carefully controlling our use of energy, new raw materials, cutting down on pollution and waste and encouraging recycling of used materials and protecting fragile eco-systems'. However, the much-abused term 'quality of life' is left undefined. It is a complex, multi-factorial variable and understanding what it means is not helped by words like 'integrating' and 'issues'. Forward Scotland, a creation of the Scottish Executive, has provided a still more elaborate definition of sustainability. Three interlocking circles provide separate compartments for each of the following: 'Jobs & Property', 'Social Inclusion & Exclusion', 'Natural & Built Environment', 'Equity', 'Sustainable Economy' and Local Environment – and then, in the shared central space, 'Sustainable Development'.[4]

Sustainability here is about dealing with the extinction of a beetle species and about analysing how green taxes affect the poor. It seems likely that the green content will fade more and more as the conceptual elasticity is tried to breaking point by including phrases like 'jobs & property', and 'social inclusion & exclusion' (?!). Searching the Scottish Office web-site with the keywords 'sustainability' and 'sustainable' turned up well over 200 items.[5] They came under the following headings amongst others: 'Sustainable transport in Aberdeen', 'Need for education about the marine environment', 'Community activities in rural areas', 'Planting trees at the Hewlett-Packard factory in North Queensferry', 'Research Programmes' for various state institutions and the 'Budget Summary'. Sustainable was often used just to mean lasting or palatable to lots of people or likely to be economically useful.

In many ways the items on the Scottish Executive's list were interesting and generally cheering. They suggested a level of awareness that did not exist as little as five years ago. One example was the 1999 *Structure Plan* in Ayrshire, an early response to the formation of one-tier local authorities and recommended for a prize because of its brevity, clarity and outreach. The plan included strong statements on promotion of the principles of sustainable development and protection, enhancement of the countryside and the environment, and so on. It sounds good, but Ayrshire's reputation is not a model one. From its filthy beaches to its problematic post-industrial recovery, that region has a large task ahead.

What does a sustainable landscape look like? It is very busy with people doing things, and, as we know, including each other into each other's activities. 'Community' and 'partnership' are keywords, as are 'consultation' and 'consensus'. The principle of – indeed, the need for – agreement precludes hard questioning of people doing what is being done. Action is motivated by negatives: the landscape is 'not preserved in aspic', is 'not a museum' and must not be 'unproductive'. I had to turn somewhere to get an idea of what the mental picture might be like and found the Scottish Executive Development Department, which is responsible for planning. The department's attitude to nature actually seems much more balanced than that of their political masters: 'The planning system has a key role to play in safeguarding landscape and wildlife' and should create 'high quality environments for living and working'.[6] Most of the colour photographs illustrating the case histories of good planning show people and/or houses, but it would be unreasonable to expect otherwise from the planning services, and the features of the schemes are tinged with green, both in the landscape and the technical sense.

Scottish Natural Heritage (SNH) are another government agency with an environmental brief: 'SNH will continue to contribute towards a more environmentally sustainable society... Our aim is to integrate natural heritage objectives alongside social and economic objectives in community plans, achieve consensus support for our vision for the future of the natural heritage and incorporate natural heritage objectives in a wide spectrum of programmes.'[7] As the latest Scottish Minister for the Environment said in a key debate: 'Sustainable development is a vast enterprise. [It] is now a key feature to our government.'[8]

Within a month of that statement, the minister had retired and his post had vanished. The ministerial responsibilities which went with that post, and included sport and culture, were shared out. The environment brief itself was carved up into sub-areas, and those without an obvious link to industry and enterprise, such as energy and water, were bundled into the already laden in-tray of the Minister for Rural Development. No doubt the balance of tasks will shift again. It is after all not unreasonable that a newly formed governmental structure should take many years to settle down into a relatively stable configuration. But during all that time when, allegedly, 'sustainability [was] the key', environment was in fact a weak area of governance. As the key policy paper *The Way Forward: Framework for Economic Development in Scotland* puts it in the brief section with the title 'Environmental Objectives' (so much for having 'environmental sustainability' built into *all* policies): 'Scotland's environment is a vital natural resource . . . A positive record of care for our country will help tourism, the attraction of inward investment and the marketing of whisky and other products in the food and drinks sector.'[9]

In this kind of atmosphere, it had attracted relatively little attention that environmental policy had been run in tandem with sport and culture. The Environment Minister was: 'responsible for the environment, renewable energy, water, natural heritage, sustainable development, strategic environmental assessments, the land-use planning system, sport, culture and the arts, the built heritage, architecture, Historic Scotland and lottery funding'.[10] A right mixed bag, you might think. The minister before that had the more reasonable, if huge, task of looking after environment and transport, but her problem was that she had not been allowed a deputy minister. Again, given the hard facts of how responsibility at the highest level has been shared out and the spending allocated, it is not surprising Scotland's landscapes have turned into commercial backdrops for whisky advertisements and tourist attractions.

Political environmental management is not a subject of much public concern in Scotland, where the green debate tends to be a reflection of its English counterpart. The Westminster environmental establishment is led by the huge Ministry of the Environment, Transport and Regions and is teeming with quangos, agencies and NGOs, with an active Green Party in the background. The Prime Minister has been vigorously criticised after some recent green speeches. It seems to me – at least very generally – that the difference in the level of debate can be illustrated by a comparison of the critique by Ian Bell in a Scottish quality Sunday paper, and an analysis by Philip Inman in a mainly English-oriented quality daily.[11] Ian Bell makes a political case against the Prime Minister, but with little detail other than what can be gleaned from skimming headlines: green taxes are a 'stealthy way of raising revenue', BSE is 'a murderous scandal' and the Food Standards Agency 'struggles at every turn to obey the wishes of American agri-business'. Maybe so; but by comparison, the equally aggressive piece by Philip Inman is a model of careful inquiry into the muddled effects of green taxes on recycling and land-fill behaviour, energy-saving policies and commercial as well as public attitudes. It would be good to see more analytical writing of this kind in Scotland – the environment staff in *The Herald* stable and the top NGOs, led by Friends of the Earth Scotland, cannot manage the entire debate on their own for much longer.

Our Need for Quangos and the Policy of Partnership

The role of quangos – quasi non-governmental organisations[12] – is difficult to define, especially for those near the borderline to real non-governmental organisations (NGOs) and voluntary/charitable bodies and trusts. Quangos can and do deal with everything and anything, overlap extensively and sometimes make decisions of overarching importance. Quangos have a statutory obligation to communicate and consult with each other, their national executive departments and the public. Some are proud about their close but specialised relationship with the Executive and the Parliament but other arms-length-from-the-state organisations can be bad-tempered about being mistaken for quangos.

Some, if not all, of the quangos are useful specialised agencies that regulate and manage funding for specific areas such as housing or nature conservation, but none seems to have won much popular support. In the run-up to the first election to the devolved Scottish Parliament, Alex Salmond and his (then) Scottish National Party campaigned for setting light to 'the bonfire of the quangos'. He had not invented this phrase; it had been judged a vote-catcher by the Labour Party in its successful national election campaign in 1997. Scottish Labour liked it too, but so far nothing much has happened, which suggests that if quangos did not exist our rulers would have to invent them. Both politicians and 'the people' presumably worry about the power of entrenched bureaucracies, or in the words of Ian Mitchell: '[Is there] a connection between idealism and authoritarianism? Are people who do not trust others to behave as they want them to, inherently prone to ruthless compulsion in their dealings with fallen humanity?'[13] A rhetorical question about his hated quango-people, but worth pondering over nonetheless, like so much in Mitchell's writings on who should control the Scottish countryside.

At present there is usually said to be about 100 Scottish quangos (sources differ on a number between thirty-eight 'executive bodies' to 460 'arms-length' organisations). Between them, they spend in the order of £6.5 billion annually. A significant proportion has at least some influence on the environment. The picture of organisations, sometimes co-operating, sometimes competing or in open conflict and all under ultimate control of the not always united Scottish Executive, is still incomplete. The next most powerful group in the world of environmental management is the non-governmental organisations (NGOs), a phrase that tends to be used synonymously with voluntary organisation (see box). Almost all are registered charities, but government money and support are not excluded. Some of the environmental NGOs are influential because of their single-minded pursuit of popular goals. Among those, the top ratings in Scotland, in terms of membership, funds and landholdings, must include the Royal Society for the Protection of Birds (RSPB), the National Trust for Scotland (NTS) and the John Muir Trust (JMT). These too attract controversy, and are if anything, subject to even more scorn and irritation than quangos by critics like Ian Mitchell. The other main group, led by Friends of the Earth Scotland, have more general goals and normally no or little fixed property.

Scottish Environment LINK's member organisations:[14] Association for the Protection of Rural Scotland; Association of Regional and Island Archaeologists; Badenoch and Strathspey Conservation Group; Biological Recording in Scotland; British Association of Nature Conservationists Scotland; British Trust for Ornithology; BTCV Scotland; Butterfly Conservation (Scotland); Cairngorms Campaign; Council for Scottish Archaeology; Friends of the Earth Scotland; Friends of Loch Lomond; Hebridean Whale and Dolphin Trust; John Muir Trust; Marine Conservation Society; Mountaineering Council of Scotland; National Trust for Scotland; North East Mountain Trust; Plantlife; Ramblers' Association Scotland; Reforesting Scotland; Royal Society for the Protection of Birds; The Saltire Society; Scottish Council for National Parks; Scottish Countryside Activities Council; Scottish Countryside Rangers Association; Scottish Field Studies Association; Scottish Native Woods; Scottish Wild Land Group; Scottish Wildlife Trust; Society of Antiquaries of Scotland; Sustrans Scotland; Wildfowl and Wetlands Trust; Woodland Trust Scotland; WWF Scotland.

Other organisations: Association for Scottish Ski Areas; Botanical Society of Scotland; British Association for Shooting and Conservation; Flying Scot; Gliding Association; Greenpeace; Heather Trust; Highlands Birchwoods; Lantra Trust; Mountain Rescue Association, regional branches; Mountaineering Bothies Association; Popular Flying Association; Royal Scottish Geographical Society; Scottish Mountaineering Club; Scottish Ornithologists Club; Scottish Scenic Trust; Snowsport Scotland; Voluntary Action, regional branches.

The inclusion of some is arguable, for instance not-quite NGOs like the Royal Geographical Society. As for the list of players in the Scottish environmental field, it includes many more stakeholders, notably the Convention of Scottish Local Authorities (CoSLA) and their autonomous members, the thirty-two Regional Councils, and the many large and small trade unions and business associations, for example the Scottish National Farmers Union (SNFU) and the Scottish Country Landowners Federation (SCLF), the Community Councils and the huge number of local interest groups.

The Scottish political scene seems to be growing larger and shrinking at the same time. Maybe it is symptomatic of 'The Third Way', that most English of concepts, which has become a byword for modern governance and has been translated into most European languages. Democracy is redefining itself, away from representative parties and regularised voting, into a varied mosaic – if not always very colourful – of interest and pressure groups. Perhaps this has become especially true of the world of environmental policy-making.

The political picture is gratifyingly (for biologists, at least) reminiscent of biological evolution at an early stage: social and political groups have been merging, dividing and budding, the offspring sharing varied amounts of the parental characteristics and expressing a creative, interactive mix of influences from past generations. There has been a parallel development towards inclusiveness and consultative processes. The proliferation of interest groups, state supported or not, might be a characteristic of a mature democracy. It has caused fascinated comment by political commentators such as Anthony Giddens: 'At an economic level, they [the people] don't believe that politicians are able to deal with the forces moving the world . . . It isn't surprising that activists should choose to put their energies into special-interest groups, since they promise what orthodox [democratic] politics seems unable to deliver.'[15] The paradox of democracy is that majority rule must be combined with regard for minorities. It has driven the development of ideas about partnerships and consensus.

From another perspective, the fragmentation of the political scene is counterproductive. Many fear that this is the case with environmental decision-making in present-day Scotland. In spite of public concern and state commitments, division has allowed rule by economic interests. An example of marginal gains and massive losses can be drawn from the debate between nature conservationists and defenders of sustainable development in the accepted political sense.

The debated landscape area is the mountain massif in the Scottish Highlands that has Cairn Gorm as its central peak. It is truly iconic for Scotland. It is also beautiful and remarkable enough to win international recognition in spite of what has been decades of careless management. The Cairngorms area is due for a National Park designation at the time of writing and is discussed in more detail in Chapter 5. Here I just want to focus on the effect of an interest group called the Cairngorms Partnership. It was set up by the government in 1994 to 'prepare and implement [by co-ordinating actions taken by the partners] a management strategy for the Cairngorms that would guarantee a sustainable future for this most special area of

Scotland.'[16] The partnership's brief in fact centres almost entirely on the local environment: protection, regeneration and 'sympathetic management of recreational pressures' in relation to the landscape of high hills and native woodlands, including deer management, moorland, agriculture and water. But it is also charged with the much less specific task of 'promoting the social and economic well-being of local communities'.

The membership is very large (see list in Chapter 6, note 16). Among the 'Primary Partners', the Scottish Executive is directly represented by the Rural Affairs Department, but is also present – by extension – in the shape of eleven quangos. These range from Historic Scotland to Scottish Homes and include not only Scottish Enterprise and Highland & Island Enterprise, but four Local Enterprise Companies (LECs). Other 'Primary Partners' range from the Aviemore Partnership, to the Scottish Landowners' Federation (SLF). The list of seventy-three 'Secondary Partners' contains almost every other conceivable organisation. In addition, the twenty-six local Community Councils are linked to the partnership. The secondary group includes voluntary and professional organisations and some more quangos, for instance the Mountaineering Bothies Association, the Botanical Society of Scotland and the Scottish Arts Council. The biggest subgroup of NGOs is of course the environmental one, which includes Scottish Environment LINK, an umbrella organisation for some forty voluntary organisations.

So, well over 100 partner organisations, at a rough count, supported by a permanent staff of about a dozen people (core Executive funding pays for the staff) have joined to manage a sparsely populated and much loved area of the Scottish countryside. The main local industries – farming, forestry, countryside sports and tourism – meet for mutual benefit in this forum, and there is much that is really positive and forward-looking about this exercise in co-operation, local democracy and inclusion. But the emphasis of the brief confirms that this was primarily a Scottish Office response to stinging criticism both from UK agencies and individuals and from European and international organisations about the 'devastation' of the Cairngorms landscape. In accordance with its management strategy, the partnership has worked hard on communication (e.g. a rangers' workshop, a survey of transport needs) and environmental education as well as the practical tasks relating to local bio-diversity plans, footpath maintenance, visitor facilities, woodland planning, black grouse-friendly plantings and suchlike. But as we shall see, the successful accomplishment of all this fades into insignificance in comparison with what is actually happening in the area by way of large-scale investments in apparently unsustainable (in the environmental sense) projects serving tourism and sport. The local communities may or may not approve in principle, but who could resist the lure of jobs?

Empowered Communities?

The Cairngorms Partnership is an example of an established community-based organisation with relatively stable funding and staff contracts. It has got environmental credentials even if it is not much in favour of radical change. This seems characteristic of community ventures: steady-as-she-goes is the likeliest course to get consensus support in both rural and urban settings. Yet it need not be so. Some creative community-driven projects that combine economic and environmental regeneration are described in Chapter 6. Such projects can serve to weld people together in joint care of a place or a region, and this is an important fact that should not get lost in an untidy mass of vague good-will and red tape.

There are many organisations at all levels – nationally, in the devolved regions, in local authorities area and more locally still – that are set up to support community initiatives, so many that even a superficial survey quickly gets out of hand. As an example I have listed as many sources of support as I have been able to find for renewable energy schemes. It is one kind of environmental industry that gets a lot of interest from local groups.

Energy Support Unit Centre for Alternative Technology; Centre for the Analysis and Demonstration of Demonstrated Energy Technologies; Network for

Alternative Technology and Technology Assessment (Open University); New and Renewable Energy Programme; National Environmental Technology Centre and Energy Technology Support Unit (Atomic Energy Authority plc); National Energy Action and the WISE Group (Scotland); Energy Action Grants Agency and its Home Energy Efficiency Scheme; The British Wind Energy Association; Development Trusts Association.

The most complex problem for community groups is how to fund their schemes, not just at the start-up point but for a secure period of years. There is an odd, uncomfortable gap between on one hand, the high political value accorded to community action, and on the other, the lack of a stable funding system that is properly geared to helping maintain staff and machinery (it is easier to fund one-off ideas like a building or a survey). The goodie-goodieness of the community vocabulary often helps to hide the problems behind worthy content.

The axiom 'Think global and act local' has been held to be self-evidently true by environmentalists for many decades. With the 'act local' part went a deep-rooted faith in the value of communities, not only in practical terms but also in spiritual. Out of such convictions grew the international eco-village movement. Scotland has got one outstandingly and unexpectedly successful example in the Findhorn Foundation on the Moray Firth.[17] There is growing support for the other, struggling or barely existing eco-communities. Sustainable Communities Network Scotland has been set up and run by a tiny team managing on remarkably generous but erratic funding, mainly from the National Lottery. The general idea is that people left to themselves, without interference 'from above', choose the right goals and work for them without being oppressive, greedy, wasteful or otherwise badly behaved. Put like this, the belief is obviously naïve, yet, the vocabulary and large parts of the belief-system have been absorbed into the current community development structures, and codified.

The *Dialogue about the Strategic Framework for Community Development* is part of a consultation exercise generated by the UK Standing Conference for Community Development (SCCD), the Community Development Foundation (CDF) and 'a number of national conferences including BASSAC and Community Matters'.[18] The *Dialogue* has led to a list of community development (CD) keywords: 'Social Justice, Participation, Equality, Learning, Collaboration' and commitments: 'Changing the Inequalities, Promoting Change, Supporting Community-led and Collective Action, Challenging Discrimination and Oppressive Practices, Encouraging Networks, Ensuring Accessibility and Influencing Policies and Programmes'. No wonder that community workers and community development managers need their 'ABCD framework' (I have not been able to find out what this is). They also need Community Development Centres (there is one in Scotland), Community Work Training Groups, 'taster and introductory' skills courses through the Open College Network, short courses for tutors, and national vocational as well as higher education qualifications.

These lists may look mischievous, but they are derived from actual documentation. There is, of course, nothing wrong with a group setting itself highly moral and carefully articulated aims, or establishing itself by setting up a common system of standards and qualifications. In this area though, there are very visible gaps between intentions and achievements – between aspirations and resources. Community projects tend to be short-term, shakily funded, under-staffed and vague in outline. Even in class acts, like the Glasgow Year of Architecture and Design, many of the community ideas were oddly insubstantial.

Meanwhile, the vocabulary of community development joined the small change of official talk and progressed to the next stage, where vocabulary actually guides legislation. One example is the concept of community ownership, a prominent part of the Scottish Executive's land reform programme, which is discussed in Chapter 6. The idea is that community ownership is good, but (it seems) that ownership by families or limited companies tends to be less good. The campaigner for land reform in Scotland, Andy Wightman, has said: 'Community involvement [in land use] begs some fundamental questions about who is going to be involved in whose land, when and with what authority, powers, responsibility and resources.'[19] As he asks later in the same section, referring to The Land Reform Act (Draft): 'Why [should busy people] be part of an elaborate 'community consultation' exercise, designed to legitimise local landed interests and satisfy the misguided political fashions of would-be land reforming politicians?' Andy Wightman does not dwell on the possibility that communities might, in one sense or another, mismanage the land as much as anybody in the absence of proper regulations, but his targeting of politically correct but empty verbiage is very much to the point.

Although much is going on, the community development scene in Scotland is not as advanced as elsewhere. The leading organisation is probably Dùthchas, which defines itself as 'Working for the sustainable development of remote rural areas of Scotland'.[20] It is sponsored by the European LIFE Programme and the Scottish Executive, allegedly through its Sustainable Development Team. As discussed elsewhere, this team is as elusive as the Executive's Environment Team. The teams may in fact consist of the same people. Both have staff members, who return tight-lipped answers to email inquiries, but the SD team have also got a rather bleak website (www.sustainable.scotland.gov.uk). Dùthchas (a Gaelic word meaning 'kinship' or 'sense of belonging and connection with the land') is an example of the increasingly familiar organisational animal, a large (twenty-one partners) partnership that looks a little like a quango with an NGO profile. It is an umbrella organisation for other groups, such as the Moray Firth Partnership, SafeinHerit Network and the TITAN offices in Norway and the Highlands.[21, 22]

In passing, it is worth pointing out that TITAN, a European IT initiative, has got a quite complex Scottish structure, focused on a joint organisational network called HI-Ways (HI stands for Highlands & Islands): 'www.hi-ways.org is a result of the collaboration of various public sector organisations and agencies in the Highlands and Islands . . . Primary providers of information and services as well as controllers of site policy are The Highland Council (TITAN Project Regional Co-ordinator),

Highlands & Islands Enterprise and Business Information Source and Northern Constabulary.' This is not enough members, of course, and all the usual suspects line up to take part: Scottish Tourist Board, Highlands of Scotland Tourist Board, Crofters Commission, Highland Council, Highlands and Islands Enterprise, Western Isles Enterprise, Skye and Lochalsh Enterprise, Caithness and Sutherland Enterprise, Scottish Homes, Scottish Environment Protection Agency, Scottish Arts Council, Rural Forum Scotland, North of Scotland Water Authority, Historic Scotland, Forest Enterprise, Forestry Commission.

Back to Dùthchas. Its vocabulary is as rich in fine feeling as any in this line of work: '[We intend to] to listen to local people, our most precious asset; to centre our energies and enthusiasm on what's best for our communities; to better understand the value of our natural surroundings; to decide on the best ways to manage our natural resources; to recognise how very distinctive our cultural identity is; to think of ways to nurture that

4. *Crofting at Melness, Sutherland.*
© Colin McPherson

identity; to strengthen the local economy; to choose the right path for development, building on our local assets; to work with various public agencies by making the most of their knowledge for our benefit; to try to convert good, sound ideas into reality.'[23] Much about its activity and its public profile is reminiscent of Forward Scotland, though with a brief more strictly confined to areas of rural deprivation. The large website is linked to its four rural 'key areas' and each lists a great deal of committee work. While objectives and visions are presented well and at length, the actual tasks started, let alone completed, seem to be mainly in the survey-awareness-raising-course-organising-cultural-tourism area.

Interview with Geoff Fagan

(Senior Lecturer, Department of Community Education at the University of Strathclyde & Project Director of CADISPA)

Dùthchas does not at any time acknowledge that other organisations are working outside its framework, with allied goals, but very much less well financed. A case in point is the CADISPA organisation, run by a dedicated member of staff at a Strathclyde University and a few colleagues.

Geoff Fagan is an academic with a mission in the 'real world' outside the university: he runs the organisation called Conservation and Development in Sparsely Populated Areas (CADISPA), set up to assist rural communities with their projects in the spirit of sustainable development. More specifically, CADISPA promotes the more sophisticated version of advocacy that sets out to teach groups of people representing local communities how to 'advocate' for themselves, that is how to deal in a competent, informed way with local authorities and other public and commercial organisations. The key idea is that ordinary people should learn to identify what they want, understand the arguments and decide on their own priorities.

The CADISPA idea grew during the 1980s out of projects aimed at engaging the young in civic and green issues, using a method called 'Critical Conversations', and took the form of an alliance between what was then the Community Studies Department at Jordanhill College of Education and the World

Wide Fund for Nature (WWF). Given its community education base, it is not surprising that there is still a bias towards keeping in touch with the young, especially through schools. Much of CADISPA's research and publications programme is focused on teaching or, 'enabling the young'.

As sustainable development became a better defined goal after the Rio Summit and *Agenda 21*, Geoff Fagan and CADISPA turned to the problem of how to work with whole communities along the lines laid down by organisations such as the Standing Conference on Community Development. The European Union began to take an interest in community-based schemes and by 1995 CADISPA had attracted international attention for its idea of providing free advisory expertise on 'soft' three-year contracts with local groups. The ideological underpinning of the CADISPA approach derives from the left-of-centre ethos seen in other community-oriented organisations, but includes a mainstream view of essentials. Its 'triangle of sustainability' is a bit shaky geometrically but sound in practice: legal validity – local and secure organisation – economically sustainable – environmentally sensitive. Apart from the strict insistence that local groups themselves tackle projects and define goals and problems, the most unusual thing about CADISPA is its link to a large, regional university.

We agreed that the university link is a precious feature, given the distance between so much of the tertiary education establishment and social development across the board, not just at 'local community' level. Geoff Fagan seemed to have overcome many of the institutional problems, in some cases by ignoring them. He felt that the most trying aspect of his outreach project was lack of time. His academic duties are as time- and energy-consuming as are his colleagues', yet in addition he has to spend hours travelling, listening and thinking in support of his local community trusts. The travelling distances are huge in this wide-meshed network, stretching from Burghead on the north-east coast to Easdale, Skye, Colonsay, Tiree and many other places in the Western Isles.

Apart from time, funding the project is a constant preoccupation. Funding worries are part of life for all academics with research programmes of their own, but for Geoff Fagan the usual problems of raising money to support his work are compounded – not assisted – by its social aims. Also, he said, the more genuinely gap-bridging a project is, the more difficult to fund. This is in spite of the many official initiatives targeting community development, from Forward Scotland's many grant options to the themed largesse of the New Opportunities Fund. The early enthusiasm for extending the CADISPA idea into a European network both worked and did not work. There are now five European branches, and some EU finance, but the branches are independent and not much of the money has found its way back to Scotland.

Funding frameworks have got to be rebuilt each time a new project is added to CADISPA's growing portfolio. The bond with the powerful WWF UK seems to be weakened and that with WWF Scotland ruptured several years ago for reasons so obscure that in spite of Geoff Fagan's frankness, I cannot extract the ins-and-outs from my notes. NGOs, like WWF, should be helpful but still have to be seen to focus on their own goals. Still, it is odd that the state is not more supportive. Even the attempt to secure a grant from the Millennium Lottery Fund fell through. It was the usual obstacle of having to raise matching monies (£210,000 in this case) from other sources. Curiously, the mere fact of having been granted the money by the National Lottery Charities Board seems to have led to sponsors turning to other, 'more needy' charities. The quangos most involved in working with the local CADISPA trusts, such as SNH and Historic Scotland, are generally supportive but tight with cash and have besides, like the Executive's Rural Affairs Department, their own fish to fry in Dùthchas. Oddest of all, in some ways at least, is the reluctance of the LECs to take any interest in what should be their top-rated spending priorities: self-help enterprises in often quite depressed regions.

However, Fagan and CADISPA are not giving up. Some of the project work – for instance the sewage and recycling centre on Easdale, the harbour rebuild on the same island, the Tiree cattle market cum community centre, the community hall on Colonsay and many, many others making a total of twenty-two projects to date – is going very well. What interests him in particular is the strong 'lifelong learning' effect of these locally driven projects. He was excited about learning for life and the whole environmental education drive that ended with such a dying fall in the late 1990s. This will be discussed in the next chapter. However, he took sides against the formal educators in the debate that followed and still felt that there should have been restructuring of formal schooling in the direction of an interdisciplinary approach. He argued that while formally educated people can provide useful ideas of how to deal with a set problem, they tend to be technical fixes rather than integrated solutions.

So, where should CADISPA be going next? Fagan speculated about a 'Third Way' type of future, that might tend to push it into the NGO sector but could also place it as a completely new style academic department, providing real and continuing community experience rather than 'sandwich courses'. Could it be that Strathclyde University would be the first environmental university in Scotland?

Interview with Kevin Dunion

(Director, Friends of the Earth Scotland)

Kevin Dunion is one of the most important and inspirational people in Scotland's environmental movement. He is very knowledgeable about international as well as Scottish conditions. His membership of the Ministerial Group on Sustainable Scotland[24] (MGSS) seems to indicate that environmental policy might become a more significant part of the Scottish Executive's activities.

He is obviously pleased by the change that the setting up of MGSS implied. Even more so, because compared to an earlier variant form, the Secretary of State for Scotland's Advisory Group on the Environment, it is more influential and has a wider remit. He cited the direct access to ministers and the right to call in statements in evidence, right up to ministerial level. The status of the MGSS membership certainly seems reassuring to the outsider and so does the weightiness of the subjects they have been discussing over the year of the group's existence. These subjects have ranged from renewable energy replacing fossil-fuels for electricity generation, to waste and energy management in large state services such as healthcare and prisons, to environmental indicators for Scotland of the type already in existence elsewhere, including in England and Wales.

But Kevin Dunion does not want anyone to become overexcited about progress so far. He argued that there is no real 'architecture for influence' in place that would allow MGSS decisions to be turned into tangible outcomes. Rather, the ministerial group is 'surrounded by a moat' that seems unbridgeable, perhaps as much due to their joint lack of conviction as to Civil Service unwillingness. What kind of structure would reinforce the policy-making process most usefully? He speculated on the idea of shaping policy jointly with the stakeholders, but is probably more aware than most of how counterproductive stakeholder participation can be. Casting his mind back to the Forum on the Environment – a make-over of the Rural Forum, now expired – and the old advisory group, both with wide-ranging memberships drawn from quangos and NGOs on the familiar partnership model, Kevin Dunion said that the consensus required to produce any decisions at all often meant diluting proposals to the point of making them meaningless.

He is prepared to take the state's regulatory options seriously, but not to commit himself to using legislation as the only framework for environmental policy. How to manage environmental policy-making is perennially difficult, and not made any easier in Scotland by the Executive's ambivalence. Both regulation and taxation can have powerful effects. Environmental taxes, from landfill tax to the carbon levy, are becoming more common. He said that the collaborative, voluntary approach has strengths as well as obvious weaknesses. The EU recognises that penalties are not sufficient to enforce regulatory provisions and in Scotland, the Scottish Environment Protection Agency (SEPA) prides itself on collaborating with industry. On the other hand, investments in the environment can look expensive in the short term and especially when local authorities and enterprise companies prioritise 'new jobs in the area'.

There is hope, Dunion feels, in the enlightened self-interest shown by a growing number of businesses. He mentioned both large pro-environmental groups such as World Business Centre for Sustainable Development (WBCSD) and smaller but all-encompassing thought-experiments such as the Swedish-founded The Natural Step (TNS).[25] Concepts which are now part of educated environmental debate, such as 'eco-efficiency' and its 'rebound' effect, tend to have their roots in these freebooting organisations. He went on to admit to having a soft spot for the US can-do environmentalist Paul Hawken,[26] with his insistence that 'environmental' and 'profitable' must not be seen as contradictory. Dunion referred to one of Hawken's views on the subject of operating market constraints on environmentally unsound trading – for instance by using the concept of pollution insurance. Corporate insurance to ward off heavy expenditure on prospective pollution clear-ups is an indirect economic instrument that bears thinking about. Protracted and unfair cases such as the one concerned with Lanarkshire chromium waste dumps (see below) might not have happened or would at least have been resolved by now, if the firm had had to take out enough insurance cover to meet future decontamination costs.

Of course, some would argue that the state has more than enough regulatory powers already, and that planning law is an obvious example, for worse as well as for better. It plays a key role in much that happens and does not happen. Kevin Dunion seems at least broadly to agree with me that much of planning law as it stands is too detailed, too vague and too beside-the-point, all at the same time. Specific instances include two continuing cases, the Cairn Gorm funicular railway and the Lingerbay quarrying proposal.

The Cairngorms case involves a (largely) local financial grouping adding to the landscape by constructing a funicular railway taking skiers and sightseers up Cairn Gorm mountain, and the Lingarbay case a foreign-based multi-national taking away a substantial part of the landscape on Harris, the Roineabhal mountain, to sell it off as crushed aggregate. There the contrasts end: both landscapes have protective designations (respectively being a World Heritage Site and a National Scenic Area), both proposals are of at best dubious economic value to Scotland (let alone to local people), both go contrary to any reasonable nature conservation policy and both have been fought bitterly in the media and, crucially, through Scotland's planning system. In the Cairn Gorm case, the financial interests won the battle but may well lose the war. The Lingarbay case has gone to appeal, mainly on technicalities. Kevin Dunion did not allow himself to comment much beyond admitting that the legal costs for his organisation are extremely hard to bear even though Friends of the Earth is a popular and powerful NGO.

Development issues such as these have their far-reaching implications both locally and internationally: the EU is investing in the order of £15 million in the Cairngorms project and the multi-national LaFarge Redland wants to export Lingarbay granite all over Europe. Dunion has argued that Scotland is the kind of small country that should attempt greater dispersion of funding to benefit community businesses, rather than focusing on export and inward investment, and so risking exploitation by more powerful players.[27] The Scottish Executive and its agencies – be they quangos such as Scottish Enterprise and its Locate in Scotland little brother or other organisations – are far too prone to swallow multi-national baits.

But it is also part of current development orthodoxy that communities should be encouraged and subsidised to run local businesses. I asked Kevin Dunion how he saw this growing trend and also its (approximate) counterpart, the buying-up of local businesses and land by absentee owners, often from outside the UK. He defended a pivotal role for the community. As for the law, he referred to the Zetland Act, which gave the Shetland local councillors – and the electors they represented – special powers to direct the income from multi-national oil exploration into the kind of developments they wanted, including their own state-of-the-art oil terminal at Sullom Voe. Importantly, the Act introduced the local community as a brake on the planning law, which is given to a presumption in favour of development. He pointed out that even the 'developer's ransom' clause is long gone; it meant that the planning authority could prescribe at least some compensation in exchange for planning permission, such as road works or a school building. 'Absentee landlordism' he argued must be counterproductive by definition, since it channelled local money away and often out of the country altogether, reduced accountability and, perhaps above all, left ownership in the hands of people with no local roots and sense of belonging.

Dunion is concerned about the role of the law, complete with enforcements and incentives, in an environmental context. As an example of what he is working on for his forthcoming book on 'environmental justice' he told me about the chromium contamination in South Lanarkshire communities of Cambuslang, Rutherglen and Toryglen.[28] The deposits are expensive to deal with, especially as the chromium has entered the groundwater to the extent that it was found leaching into the River Clyde in the early '90s. They were dumped by the Rutherglen chemical works, long since defunct. The hot potato of decontamination has been passed for years between the local action group, their MP, the Greater Glasgow Health Board, Strathclyde (then) and Glasgow (now) Councils and the Secretary of State Office (then) and Scottish Executive (now). In 1999, the Scottish Office itself officially acknowledged that 'there are significant areas of contaminated land along the line of the proposed [M74 extension] scheme that are a result of previous chemical works dumping chromium waste over a large area'.[29] Who should act and what should be done is far from clear in law. But then, Dunion said, pleas that 'something must be done' were weakened as a curious intervention by scientists from Glasgow University, commissioned to evaluate the risk to local people, found there were no health deficits linked to the waste, that in spite of its extent and acknowledged toxicity.

Dunion is concerned about the role scientists play in the understanding and management of the environment. His statement to the inquiry of the Scottish Parliament Environment Committee into the advisability – or otherwise – of releases of genetically modified organisms into the environment is a case in point. His written statement includes both carefully argued questioning of the design of the trials and arguments in favour of recognition of public interest, social and ethical issues. The kind of science he dislikes was exemplified in his story from the 1999 Edinburgh Conference on the use of GMOs: anxious questions from the gallery, however well-informed, were received with peremptory demands for evidence 'from peer-reviewed papers in well-established journals'. Since GMOs have been sprung on the public and there are few such papers anyway (especially few critical ones), this silenced, though did not calm, the opposition.

We agreed with some enthusiasm that the take-it-or-leave-it science presented to the public by bodies such as the Food Standards Agency (FSA) – specifically set up to deal with scientific investigations into food safety – is completely out of order. Anyone who has tried to extract anything of actual scientific worth (let alone practical advice) from the FSA staff dealing with the potential dangers of eating farmed salmon will know the reason for this harsh judgement of an apparently decent quango.

In the end, the landscape became part of the conversation. The 'environment' is not the same thing as the 'landscape', but the two sets of concepts are deeply related. We talked about it briefly and Kevin Dunion referred to the debate in the ministerial group on renewable energy and the problem of where to site the (Danish-made, Dutch-owned . . .) wind generator plant. It was not hard to get agreement, because Scotland needs renewable energy and windfarms can be beautiful in their own right. He then went on tell me about how the QCs acting for the parties in the Lingarbay case spent a couple of days arguing about landscape aesthetics. It seems absurdity was the order of the day. I asked for more information and Kevin Dunion supplied a long and fascinatingly 'expert' extract ('Quarrying would not be visible from most of Rodel for the great majority of the working period'), which will be described in more detail in Chapter 10.

The 'expert' is a figure who stalks every byway of environmental policy debate. As the sense of impending doom becomes greater, expert advice is referred to with increasing frequency and ever more blinkered faith. It is a function of all our lack of understanding that 'what the scientists say' has become such an uncritically used argument.

Notes

1. The Cost-Effective Landscape. www.scotland.gov.uk/library/docs-l/ldpart1 'Review of Existing Methods of Landscape Assessment and Evaluation'
2. by the Macaulay Land Use Research Institute, Robert Gordon University, and Leicester University, in *Cumulative Impact of Wind Turbines*. bamboo.mluri.sari.ac.uk/ccw/task-two/evaluate from the section Subjectivity versus Objectivity.
3. *Scotland's 21 Steps to Sustainability* CoSLA , 2000. See also Sustainability in Action Practice Guide www.cosla.gov.uk/index.asp *Agenda 21* is a global action plan, designed as an all-round guide to environmentally sound management for local government; see also Note 16, Chapter 1.
4. *Towards a Sustainable Future. Annual Review 1999, Forward Scotland*, p. 9. Cf. also www.transformscotland.org.uk
5. www.scotland.gov.uk
6. *Planning for Natural Heritage. Planning Advice Notice (PAN) 60*. Scottish Executive Development Department, 2000, p. 39.
7. Operational Plan 2000/2001: Programme 8 – Supporting Sustainable Development. Scottish Natural Heritage www.snh.org.uk/about/OpPlans/op-10.htm
8. Sustainable Development debate introduced by Sam Galbraith. *Scottish Parliament Official Report*, 11 (01), col. 20, 28 February 2001.
9. The Way Forward: Framework for the Economic Development in Scotland, Executive Summary, under the heading 'The Focus of the Intermediate or Enabling Objectives' www.scotland.gov.uk/library3/economic/feds-02.asp
10. Ministerial Responsibilities. www.scotland.gov.uk/who/ministers.asp. Note that the definitions have changed since the text was written; as of 1 June 2001, the new-style Minister for the Environment and Rural Development is: 'responsible for policy in relation to the environment and to rural development including agriculture, fisheries and Forestry'.
11. 'He thinks blue . . . and we see red' by Ian Bell. *Sunday Herald*, 29 October 2000; 'Green at heart? What a load of old rubbish' by Philip Inman. *The Guardian*, 13 January 2001.
12. Quasi non-governmental organisation; later glossed as Quasi-autonomous non-governmental organisation, and then renamed again as non-departmental public body (it did not catch on) = a semi-public administrative body outside the Civil Service but with financial support from and senior members appointed by the government (quangocrats).
13. *Isles of the West. A Hebridean Voyage* by Ian Mitchell, 1999, p. 220.
14. Member Organisations www.scotlink.org.

15. *Runaway World. How Globalisation is Reshaping our Lives* by Anthony Giddens, 1999, p. 74.
16. The Cairngorms Partnership. What We Do. www.cairngorms.co.uk
17. www.findhorn.org See also www.ecovillages.org/scotland/scns/AboutSCNS Sustainable Communities Network Scotland (SCNS) is a charity set up in June 1998 which promotes the provision of education to the general public on sustainable development. SCNS is currently working with six groups, to develop plans for sustainable community projects using the eco-village and cohousing models developed in Scandinavia and North America.
18. A Dialogue about the Strategic Framework for Community Development and The SCCD Charter including A Working Statement on Community Development by Standing Conference for Community Development. http://homepages.nildram.co.uk/~sccd/
19. *Scotland: Land and Power. The Agenda for Land Reform* by Andy Wightman, 1999, p. 79.
20. What is the Point of Dùthchas? www.duthchas.org.uk
21. The SafeinHerit Network is a partnership set up in 1999 and made up of communities from Scotland, north Norway and Swedish Lapland. It was a response to a European Union/Norwegian Government initiative to stimulate community development and environmental protection in northern parts of Europe through the Northern Periphery Programme (NPP).
22. TITAN is a pan-European consortium of Regional Service Providers and Industrial & National Public Network Operators. The idea is to provide IT links to help initiate 'regional and trans-national service integration activities', which might 'improve the quality of life for citizens and Small & Medium sized Enterprises (SMEs) in rural areas'. The key goal is to provide advanced telematics tools to enable efficient navigation to public information and interactive services, of use as business support, local and public service provision, lifelong learning, and community networking. Four European centres are involved, including one in Norway and one in the Highlands.
23. What is the Point of Dùthchas? www.duthchas.org.uk
24. The MGSS first met in January 2000. It is chaired by the Minister for Environment, Sport and Culture and its members are the Ministers for Transport, Finance and Local Government, Social Justice and the Deputy Minister for Enterprise and Lifelong Learning and Gaelic. It is the only ministerial committee of its kind to have two external members: Kevin Dunion (Director, Friends of the Earth Scotland) and Mark Hope (Director, External Affairs at Shell Expro). The group 'works with Ministerial colleagues to integrate the principles of environmentally and socially sustainable development into all Government policies; to take a strategic approach to environmental issues; to work with people in Scotland to bring understanding that sustainable development offers

benefits now and protects them in the future'. Information can be found at www.scotland.gov.uk

25. WBCSD is a 'coalition' of 150 companies from twenty plus major industrial sectors and some thirty countries. TNS is a non-profit-making organisation for integrating science, environmental and business principles.
26. Paul Hawken is a 'leading architect and proponent of corporate reform with respect to ecological practices', a dynamic director of many companies of his own and a co-writer with Amory and Hunter Lovins, of Factor Four fame (see Chapter 8).
27. 'On the Scottish Road to Sustainability' by Kevin Dunion in *Scotland: The Challenge of Devolution* by Alex Wright (ed.), 2000.
28. All forms of chromium are toxic in humans if ingested/ inhaled in large enough concentrations, and some compounds formed with hexa-valent chromium (chromium(VI)) are classed as carcinogens. Chromium(VI)-containing waste is a pollutant in law and the onus is on the polluter to deal with contaminated land and water.
29. The Strategic Roads Review, November 1999. www.scotland.gov.uk/travelchoices/docs/tsrr-02

4
Education and the Environment

Education about What?

THERE IS AN AMUSING and acute passage in Noreena Hertz's book about global capitalism describing the vigilant daily life of the aware consumer.[1] It starts: 'Open my bottle of Ecover and squeeze biodegradable liquid over yesterday's plates crusted with residues of GM-free organic pizza. Fill cafetière with Fairtrade . . .' On and on goes the list of responsible products: the CFC-free hairspray, the not-tested-on-animals cosmetics, the credit card issued by an ethically aware investment trust, the unleaded low-sulphur petrol from a politically acceptable oil company, the snacking on Ben & Jerry's ice cream. What, you did not know either why it is an environmentally approved act to eat Ben & Jerry's? I learnt that their slogan is 'We don't cut down trees in the Amazon'.

All these decisions are desperately serious. You do not learn to make them wisely in the great university of life. Each one has been researched, argued over and put into context by workers in many academic disciplines. It is up to global citizens to know enough about the facts and processes to understand the conclusions. It is just tough that the areas range from atmospheric chemistry to nutritional physiology and animal behaviour to international trade and politics. It is even tougher that education is such a double-edged sword to wield in this daily battle to do the right thing. Environmental education, as we shall see, is no easy matter. Currently it is being complicated further by precisely the processes of global capitalism that are the subject of so many recent articles and books, including the one by Noreena Hertz. For instance, both Naomi Klein in *No*

Logo and George Monbiot in *Captive State* present evidence to show that from primary school to research institute, education is being subjected to creeping takeover by commercial interests.[2] George Monbiot lists hair-raising examples from British schools (e.g. science, courtesy of British Nuclear Fuels: 'Accidents happen all the time. Can you think of any accidents that have happened to you . . . ?') and quotes from the European Round Table of Industrialists: 'Too often, the education process itself is entrusted to people who appear to have no dialogue with, nor understanding of, industry and the path of progress . . .'

In this atmosphere, education must be by definition highly politicised. Before going on to describe in more detail the current ups and downs of environmental education policy in Scotland, I have summarised an interview with Robin Harper – teacher, committed environmentalist and politician, and Scotland's first Green Party MSP.

Interview with Robin Harper

Apart from one Highland councillor (but that's another story), Robin Harper is the first elected representative of the Scottish Green Party at national or regional level. His success pleased not just 'Greens', but a wider circle of personal supporters of this likeable and articulate man.

His task as the Parliament's all-round environmental conscience must be harrowingly difficult. But then, Robin Harper is used to working on his own. He spoke of how he 'beat the system' and won his Lothian constituency: it took years of personal involvement. It is the same for all who want to get elected, he said, only more so for candidates without powerful party machines. Harper, already rooted in the community as a school teacher and extra-curricular music teacher, among other things served on the Lothian Health Council, the Lothian Children's Panel and as the president of the local EIS.[3]

He feels strongly about the positive value of community life and participatory democracy. The Scottish Green Party has a history of political engagement across a wide front in spite of its small membership and Robin Harper has been working hard not to be seen as 'the token Green'. He uses the cross-party group system of the Scottish Parliament to establish a broad left-of-centre presence. Given his previous interests, it is not hard to see why he has joined cross-party groups dealing with health and young people's issues. But even when subjects such as crofting, and animal welfare are categorised as 'green', the social groups on his list easily outnumber all others among his twenty-two memberships (twelve as convenor or vice-convenor). The Parliament is not an easy place to work in, for all its structured activities and procedures. When it comes to the more interesting and powerful parliamentary committees, memberships have to be applied for and sometimes fought for. He is a member of the Transport and Environment Committee and the T&E observer on the European Committee. This is a useful role in many ways, he said, but particularly because so much of the drive for Scottish environmental regulation comes from EU directives. At the time, he felt good progress was being made on the 6th European Action Plan on the Environment. The Minister for the Environment has changed twice since then, but several of the issues that were in the forefront are still important and some – CAP modification, farming and agri-environmental schemes – are particularly close to Robin Harper's heart.

Each MSP is 'allowed' a Bill during the effectively four-year period of a Parliament and in Robin Harper's case it is a Bill demanding specific targets for organic farming and food production in Scotland. The basic premises concern the amount of farmland certified as 'organic' and the volume of organic foodstuffs produced and marketed in Scotland.[4] There are still precise numbers to be agreed on, and there will also be some debate about the target dates. The ministerial plan, to be based on the Bill if it succeeds, should also include provisions for making organic food widely accessible and promote a local and regional food economy. It has progressed well, attracted over forty signatories and is likely to have received a tremendous boost by the farming crisis at the beginning of 2001.

It s worrying that crises seem to be necessary agents in Scotland's greening process. Robin Harper ascribed the tendency for inquiries and radical proposals to disappear at least in part to the ex-Scottish Office Civil Service. His advocacy of organic food production had met with evasion by officials focused on meeting EU directives. 'We don't have to do that,' was apparently a not uncommon response. He went on to describe the management of the landfill tax credits in terms of a classic example of how a basic idea – charge for dumped waste and return the money in funding for recycling projects – has become weakened by over-complication. In the first place, Harper argued, the money should have stayed in Scotland, instead of disappearing into the Treasury coffers in Westminster. At present the credits, essentially reductions in tax for landfill waste operators who support recycling and other green schemes, have to be reclaimed from the Treasury. In fact, a relatively small proportion is reclaimed and Forward Scotland, the state agency charged with managing the scheme, is not doing so well.

Nor is household recycling doing well, nor attempts to reduce car use, nor getting customers to avoid peat-based garden compost nor excessive wrapping and plastic bag use, nor many other benign ideas that rely on participation by the public. Why are the Scottish so profoundly uninterested in environmental issues? Robin Harper felt that a significant proportion of the blame has got to be put at the door of the Scottish education system. In spite of the efforts of individual schools, departments or teachers, the topics and approaches grouped under environmental education are not taught.

Schools feel that getting their pupils through Standard Grades and Highers, as well as managing bureaucratic demands, puts too much pressure on teachers and headmasters as it is. It makes them resistant to taking on what is widely regarded as just a nice but fanciful extra burden – another civic topic of the kind that politicians dream up from time to time. It is certainly not true, in spite of some assurances to the contrary, that environmental issues form part of all subjects. Biology may include some hours dealing with the basic principles of ecology, but even such options disappear for the over-sixteens. Only in primary schools does 'the environment' feature as a significant part of the curriculum and apparently almost all the invitations to the education staff at organisations such as SNH and SEPA come from primary schools. We agreed that the excitement about environmental teaching for very young children might – perversely – enforce the lack of interest later. Members of the 12–17 age group realise that at secondary school, messing around with 'nature studies' is a thing for kiddies and settle down into blank ignorance that the Highers curriculum does little to disperse. It seems that young people in Scotland know little about environmental issues and, predictably, care less.

5. *The West Highland Way, north of Milngavie, Dunbartonshire.*
© Colin McPherson

The Rhetoric and Reality of Environmental Education

Environmental Education (EE) was not seen by its original creators as a combination of likely subjects – say, biology, geography, sociology and law. In the minds of its chief spokesman, at least until a few years ago, EE was something new, special and exciting. Rather than a group of subjects, it was a holistic concept that should be within reach of everyone and taught everywhere, as part of everything. Apart from schools and higher educational institutions, EE should be on the agenda in community centres and study circles and be part of apprenticeships, vocational training and lifelong learning:[5]

> The Report treated environmental education as a whole and not merely the sum of its parts, seeing it permeating the whole curriculum [and] not as a separate subject. It identified empirical, synoptic, aesthetic and ethical elements and showed how these may be developed through all the stages of education and different curricular subjects, urging a co-ordinated total programme, extending into informal education and later life.

But the consensus seems to be that in spite of all this idealism EE is not found anywhere in the Scottish education system. What happened? The interview with one of its creators and chief proponents, Professor John Smyth, will help to answer that question (see below).

EE had its origins in the 1970s Scottish educational establishment and reached a high point in 1993 with the publication of John Smyth's *Learning for Life*.[6] The creation of the Scottish Environmental Education Council (SEEC), presided over by John Smyth, was another apparently significant event. But at the start of 2000, SEEC had gone into terminal decline, after producing its last report *Learning to Sustain*.[7] Both it and *Learning for Life* were soon superseded by several other Scottish Office reports called things like *Scotland the Sustainable? The Learning Process* and *No Small Change*.[8] One clue to the dropping of the SEEC is found in the view expressed by the (in 1998) Minister for the Environment, Lord Sewel, who said 'The SEEC is to be congratulated . . . but now it is up to the rest of us, through our own actions, to ensure that it is no longer needed.'[9]

Nothing in the minister's speech suggested other than complete conviction that soon there would be no need to worry any more. He set up a Scottish Office-based umbrella organisation called Sustainable Scotland and left EE with its members. It promptly formed an education group (SSEG). SSEG closed down shop just over a year later, but did have time to publish a handful of reports. One – called *Down to Earth* – was intended as a follow-up to *Learning for Life*, but by then the official view of EE had become uncompromising.[10] The 'mission statement' came down in favour of leaving the education system to get on with its main task of helping pupils pass exams, as before. 'The environment' got a passing reference in a scheme to recycle computers into schools. Sustainable Scotland now seems to have receded into a near-virtual existence as a group of people servicing a web-site.

When the Scottish Environment Protection Agency (SEPA) started operating in 1996, it initially took the EE concept on board. Its 1998 *Strategy* explained the idea and emphasised that 'the word "education" has a tendency to produce an image of formal school teaching, whereas . . . it is in this wider, whole-life sense that SEPA uses the word.'[11] As late as August 2000, SEPA still spoke of 'the central role education . . . will play in the achievement of [SEPA's] objectives to protect and improve Scotland's environment and contribute to sustainable development'.[12] But by the time its Corporate Plan 2000/2001 came out, education had all but vanished from the agenda.[13]

SEPA's educational role was always elusive. Although the agency states rather grandly that it is 'responsible for environmental improvement in Scotland', more specifically its main tasks are to regulate pollution in land, air and water, and the handling of radioactive materials. The agency employs about 500 staff under the headings of Pollution Prevention & Control and Scientific Support and half as many again for running the day-to-day administration and so on, including forty presumably senior people for Policy Development. Apart from one dedicated Education Officer, some doubts remain about who should have an educational brief (see interview below).

In 1999, SEPA was still settling into its single-agency role. It did not have a straight run. SEPA is usually first in the firing line when some filth escapes into the environment or when polluters get away with a risible fine or no prosecution at all. Lack of funding for its primary role of controlling pollution has been a constant complaint. The enthusiasm for the EE idea with its vision of a populated landscape, benign and charmingly full of children and adults caring, had already faded. By contrast, SEPA's landscape is a running sore and its terminology is clinical: 'sustainable urban drainage . . . habitat enhancement . . . catchment management plans . . . minimising or eliminating risks . . . control of major accident hazards'.[14] In official SEPA-language the elements, air, earth and water are known as 'three environmental media' and education as an 'environmental strategy tool'. There is no reference in the plan to the use of this tool beyond listing it in the Introduction, not even on the single page of rather opaque comments on research and development.

Where else might one find EE in reality? Not in the universities or other Scottish tertiary educational institutions, although some have created schools of study or at least new subjects with 'Environmental' as part of the title. As I discuss in more detail later, in Scotland environmental orientation and skill acquisition at this level is far behind comparable European countries. In Scandinavia, practically every branch of higher education offers whole courses as well as course units in subjects with well defined environmental aims, and the larger tertiary institutions usually have at least one institute devoted to specialities from environmental geography to environmental product assessment.[15] At the level of secondary education, the Scottish curriculum suggests that environmental studies does not get much attention anywhere except as one of the topics under the general heading of Modern Studies, and that EE, in the idealistic sense, simply does not exist at all. Everyone mentions (see interviews in this chapter) primary schools as the exception and places where all manner of nice nature study things can be, and often are, fitted into the subject area 'science'.

But EE was meant to be – should be – much more than nature studies. Is EE part of apprenticeship training? Study circles? Not really, but initiatives at the level of community can be quite cheering. One example is the Cairngorms Partnership's publication *Learning for a Sustainable Future*.[16] For anyone who hoped EE would permeate society by the start of the 21st century it might make somewhat saddening reading. Still, it refers to many practical initiatives. Examples include an exercise in agricultural waste disposal, a meeting of community councillors on Cairngorms National Park and a Scottish Mountaineering Council booklet on how to manage defecation when out in the hills. The information pack about the Glenlivet Estate looks good, but its claim to be a 'case study of sustainable land use' rings just a little hollow for observers of the Crown Estate. For the ubiquitous gap-toothed primary pupils, the options are shredding post-Christmas trees for mulch in Grantown, nature study groups in the glens of Angus and a 'school waste minimisation project' in Strathspey.

This kind of community-based enterprise is important but looks forlorn in the context of the weak environmental teaching in the 'straight' education system. Findings such as 'The sample [of the Scottish public] showed an almost total lack of understanding of global environmental issues' ring only too true.[17]

So, why did 'classical' EE, which was practically invented in Scotland, not only disappear but remain unreplaced by anything else that works across a broad enough front? Perhaps *Learning for Life* was too dauntingly broad in its provisions: the list of 'things to do' for the Secretary of State ran into twelve complex recommendations, from 'Make a Statement of Intent on a national policy' to 'Ensure that additional costs involved in practical work . . . are taken into account when funding levels are set.' These were not enough, for to bring everybody else into line there were nineteen recommendations for 'various implementing bodies' and a whole host of additional comments, demands and suggestions for everybody.

Learning for Life is a vision expressed through the means of a Secretary of State Working Group, not usually the stuff dreams are made of. The members of the Working Group must have imagined a remarkable response: a reawakened Civil Service would move swiftly to catch up with their political masters' genuine willingness to invest seriously in a holistic approach to environmental education. Their dream landscape, apart from being populated with charming children, keen adults and comradely teachers, also contained well-informed, enlightened members of the official establishment. It did not work out like that.

A less appealing but more effective start might be made by focusing on tertiary education and strengthening the environmental remit of institutions working in land use, forestry, agriculture, veterinary science and fishery. Reforming the university-level curricula has worked in many countries of comparable size to Scotland, and not just by creating a supply of trained people to do teaching and research. Leadership from the top also has the effect of attracting clever young people into environmentally oriented courses at secondary level.

I think there are many things about this suggestion that might have irritated the original EE enthusiasts, but most of all perhaps the idea of separate courses. They were right at the time to reject compartmentalised environmental teaching. The main, hands-on reason seems to have been that such courses did not lead to good employment opportunities. The situation has changed. It is obvious that Scotland needs environmentally informed expertise here and now. The idea of the tertiary sector taking a lead is neither new nor untried in Britain. The London-based organisation Forum for the Future (FoF) runs an overlapping group of projects in higher education with interesting publications such as *Greening the Design Curriculum* and *The Engineer in the 21st Century*.[18] It also works with the Westminster Environment Department (DETR) and twenty-five universities in England and Wales on so called Best Practice Projects. FoF has been very 'forward' and even picked up grant funding for green course development from the Scottish Higher Education Funding Council. Rather than waiting for a pastoral state of EE 'permeating' all standard subjects in Scotland, it would be worthwhile setting up at least one 'environmental university' on the Dutch or Swedish model. After all, it looks as if Scotland already has a prime institutional candidate or two hovering in the wings.

Interview with Professor John C. Smyth, OBE
(ex-Chairman, *Learning for Life* Working Group at the Scottish Office)

Professor Smyth has shaped his retirement into an enviable mixture of fascinating work and foreign travel related to his lifelong interest in environmental education, and an idyllic existence shared with his immediate family. We talked on the growth of the EE idea in Scotland and the circumstances that led to its being dropped from the Executive's agenda.

The introduction to the report by the *Learning for Life* working group ended with 'If you think education is expensive, try ignorance.'[19] I feel this particular truism cannot be repeated often enough and asked John Smyth if he thought that there was enough environmental education in schools and universities to lessen at least some of the most costly effects of our ignorance about the environment. His reply can be summarised as 'Maybe in part – but not really.' Referring to the roots of the EE idea among Scottish educational theorists in the early 1970s, who were 'well ahead of the field', he spoke of the widespread enthusiasm at the time for this area of study, including the influential HM Inspectors of Schools in Scotland. The EE concept won international approval and was defined at a World Conservation Union (IUCN) meeting.

It is alarmingly easy to forget what has happened in the near past and chastening to listen to the story told by someone who was there. John Smyth listed the main official responses, which caused the break-through of the EE idea to end in its gradual break-down: 'too wide-ranging' – 'too great penetration expected' – 'premature' – 'better wait for reorganisation of Government departments and/or the Munn and Dunning report on teaching in Scottish schools'. The Munn and Dunning report in fact paid little attention to 'philosophy', even though James Munn, one of the co-authors, was supportive of the EE idea. The Scottish universities were not interested, at a time when many tertiary institutions in northern Europe were taking aspects of the EE message to heart. John Smyth said that he felt the Scottish universities had always been 'boxed-in', that is not only out of touch with the rest of education, but unable to see the point of integration. On the other hand, the further education system had been interested in EE but was too underfunded to drive any real change.

John Smyth went on to point out that a group called Learning for Life still exists at Moray House and is chaired by its principal. The Scottish Environmental Education Council (SEEC), born in 1977, closed its doors in 2000. Some of its offspring continue the good work, including the Education 21 Scotland Forum. EE modules have been designed for the Bachelor of Education curriculum and some of the larger teacher training colleges in Scotland (Moray House, Jordanhill) have taken these modules on board.

Teachers of the younger children (5–14 years of age) have a lot of 'environmental' material to choose from. John Smyth mentioned several centres of innovation but felt that even so, only the youngest children are likely to get the time and real intellectual freedom to puzzle together a personal view of a large, multi-factorial area. A reductionist 'scientific' view tends to dominate all too soon and by the time pupils get close to major exams, freedom to think is curtailed still more. As a consequence, there is little funding for projects aimed at developing more radical EE courses. Interestingly, there is more going on outside formal education. Institutions such as the Royal Zoological Society produce excellent proposals, but again have been short-changed when it came to trials and reporting.

John Smyth remains very much involved in international EE developments. He worked with the group compiling the EE section (Chapter 36) of *Agenda 21*,[20] which he says mentions the word 'education' more often than any other except 'government'. Although practically all countries are signatories of *Agenda 21*, UNESCO has recently gone on record stating that education is 'the forgotten priority of Rio'. John Smyth maintains regular contacts with the great EE 'parent' agencies World Conservation Union (IUCN), UNESCO and UNCED and has worked on a UNESCO-funded evaluation of EE programmes. In spite of honourable exceptions, most countries still operate on the basis that the most academically able pupils should learn the traditional way. He commented acidly on the interdepartmental jealousies that have everywhere got in the way of 'joined-up government' between Environment and Education Departments.

True enough, at the Scottish Office, the Environment Department had supported his report *Learning for Life* much more enthusiastically than had the Education Department, but a change in key personnel had been 'the kiss of death'. The story that followed was instructive. It was about conflicts between departmental and personal agendas, and between strong and weak personalities, as the hot potato of EE was handed from one group to the next. Somewhere along the line, the idea lost its impetus and became enshrined in anodyne reports.

Responding to the suggestion that part of the trouble was the generality of the concepts in *Learning for Life*, John Smyth granted that by being so all-encompassing, EE – like many other environmental issues – was all the more likely to become nobody's specific property. However, he felt that there had been many quite specific points made in the report. It is hard to see where innovation on the scale John Smyth would like to see is going to come from now. He argued that new ideas need to be injected into the system and that EE itself has evolved, from a jigsaw of pieces from many disciplines, via a political phase when current affairs and issues redrew the educational map, to the current view – which he shares – that a systems-based approach is the right way to learn about the environment.

Interview with Mark Wells
(Education Officer, SEPA)

Mark Wells is a former Chief Executive of the Scottish Environmental Education Council (SEEC) and is an active member of Scottish Environmental Design Association (SEDA). He is a much younger man than John Smyth but as concerned about 'classical' EE.

Mark Wells came to the Scottish Environment Protection Agency (SEPA) steeped in the *Learning for Life* premises for environmental education that it had been the task of SEEC among others, to turn into real curricula. The demise of EE concerned him a great deal. His sense of frustration was mainly directed towards a system that had come to rely too much on consensus. He was full of praise for the environmental NGOs. Like John Smyth, he admired the work of organisations like zoos and botanical gardens, but singled out the educational work 'behind the scenes' of, for instance, the World Wide Fund for Nature and the Royal Society for the Protection of Birds.

Their educational briefs are clear and backed by funds that are already in place. He felt that this compares with the most advanced – in EE terms – sector of tertiary education, the Further Education (FE) colleges. However, although they try to widen their EE remit, lack of funds has meant that relatively little is actually being done.

Mark Wells felt that the present Scottish Lib/Lab coalition, like the New Labour regime in England and Wales, has missed out on the environment in spite of prioritising education across the board. The Scottish Education Bill (2000) was preceded by a large consultation exercise but, as Mark Wells pointed out, does not even include the words 'environmental education'. But then, neither did the influential Garrick Committee's recommendations for *Higher Education for the 21st Century*.

So what does Mark Wells do? What is it like for him inside the organisation with the biggest stick when it comes to enforcing environmental good behaviour in Scotland? He replied that contrary to what might be expected, SEPA (unlike Scotland's other environmental quango SNH) does not have a statutory education duty. Both organisations recognise an 'operational duty to educate', which does not exactly help to push education budgets to the top of the list.

At this point I asked a little inconsequentially about something that had bothered me for a long time: why does the environment get such poor coverage in state-funded media advertising compared with, say, public health topics? In response, Mark Wells spoke with some feeling about the lack of creative input, precisely matching the lack of funds. The Health Education Board for Scotland (HEBS) has an annual advertising and public relations budget in the order of £4 million. The *entire* annual funding for SEPA amounts to about £30 million, of which perhaps £1.5 million (5 per cent) is allocated to broad education, information and advisory functions.

This does not sound good, but not so very bad either. How does SEPA carry out its educational work? A lot is meant to be done by the agency's staff in general and the Environmental Protection Officers (EPOs) in particular. Given that there are about 300 EPOs and limited scope for training or even achieving a uniform approach, the result tends to be kaleidoscopic. As far as the EPOs are concerned, their mainstay is the Pollution Prevention Guidelines. These suggest, but do not prescribe, how to prevent pollution, and to some extent, the methodology.

EPOs are also expected to educate in other contexts, such as 'waste minimisation programmes' and 'habitat enhancement initiatives'. I looked at a series of leaflets produced in-house, covering a wide variety of subjects, including managing agricultural sludge, dealing with waste-water run-off (from hard ground surfaces) and river-side habitats. The leaflets were nice, but showed the unmistakable signs of both budget constraints and limited educational reach. The latter is the more worrying. One feels that the EPOs should carry their expertise proudly and not simplify for people who might genuinely want to know. But then again, according to Mark Wells, some find the educational approach a bridge too far. In areas such as construction and agriculture, where regulations can be as complex as they are arcane, to 'know the regs' is usually felt to be enough.

Another unexpected problem with SEPA's educational role is paradoxically caused by its sheer authority. SEPA staff must be careful not to be seen to come down on one side rather than the other, and certainly not before decisions have been passed down 'through the proper channels'. SEPA's every move is also keenly watched by its old comrades-in-arms, the local authorities, who have retained a powerful battery of powers over planning and development.

SEPA and consumer issues are a potential minefield. Unlike many European countries, eco-labelling has never quite penetrated the UK market-place. Even little British lions stamped on eggs are meaningless and the enormous range of 'green' product labels are recognised as sales devices rather than statements of fact. There are examples of the important double role of regulator and educator in this area being handled by the Environment Agency in England and Wales. However, for SEPA, this is still a 'no-go area', as are vehicle testing to environmental standards and traffic management issues. We ended this conversation on some kind of dying fall, talking about SEPA's oddly non-environment-friendly headquarters building and the almost certainly unsustainably-grown tropical forest wood used to make the table we were sitting at.

Notes

1. *The Silent Takeover: Global Capitalism and the Death of Democracy* by Noreena Hertz, 2001, p. 119.
2. *No Logo. (No Space. No Choice. No Jobs)* by Naomi Klein, 2000; *Captive State: The Corporate Takeover of Britain* by George Monbiot, 2000, pp. 333–4.
3. Educational Institute for Scotland (EIS) is the dominant school teachers' trade union in Scotland.
4. Organic Food and Farming Targets Plans (Scotland) Bill (Draft 2). Scottish Parliament, proposed by Robin Harper, MSP (Green Party), 2001.
5. HM Inspectors of Schools in Scotland Report, 1974, summarised in *Learning for Living*, Scottish Environmental Education Council 1985, p. 5.
6. *Learning for Life. A National Strategy for Environmental Education in Scotland.* The Scottish Office, 1993. Professor Smyth chaired the Secretary of State's Working group on Environmental Education and Learning for Life was its final report.
7. *Learning to Sustain.* Scottish Environmental Education Council, 1998.
8. *Scotland the Sustainable? The Learning Process*; *No Small Change.* Both by the Scottish Office Education for Sustainable Development Group, 1999.
9. Greater Environmental Awareness Neeeded. www.scotland.gov.uk/news/releas98_1
10. *Down to Earth. A Scottish Perspective on Sustainable Development.* Scottish Office, Sustainable Development Team, 1998, pp. 9–10 (Education).
11. SEPA was formed under the Environment Act (1995) from existing agencies (England & Wales have its own EPA). A non-departmental public body with wide responsibilities, including setting environmental emission standards, inspecting, advising and when necessary, taking polluters and other offenders to court. The document referred to is *Environmental Strategy.* Scottish Environment Protection Agency, 1998, pp. 29–31 (Education and Influencing).
12. SEPA & Education. www.sepa.org.uk/education/
13. *Corporate Plan 2000/2001.* Scottish Environment Protection Agency, 2000.
14. Ibid., p. 52.
15. Miljöutbilding [Environmental Education]. www.hh.se/dep.utb/
16. *Learning for a Sustainable Future. An Education for Sustainable Development Strategy for the Cairngorms.* Cairngorms Partnership 2000, pp. 7–14 (Examples).
17. *Environmental Education: Is the Message Getting Through?* (The Scottish Office Central Research Unit, HMSO, 1994.

18. www.forumforthefuture.org.uk/new_website Forum for the Future was set up in 1996 by the well-established Green trio Jonathon Porritt, Paula Ekins and Sara Parkins and specialises in partnership projects in sustainable development.
19. John Smyth wrote (personal communication): 'I first heard [this phrase] from Al Baez, former chair of IUCN's Education Commission, who got it in turn from the Principal of Princeton University. It was picked up and used by the chairman of IUCN's NW Europe Committee at its conference in Stirling in the presence of the Minister (Lord James Douglas Hamilton), from where it was passed on by officials and used by the Secretary of State at a conference of industrialists. Beyond that I do not know its origins.'
20. *Agenda 21* is a global action plan, designed as an all round guide to environmentally sound management for local government. It was agreed at the 1992 Earth Summit in Rio de Janeiro; see also Note 16, Chapter 1.

5

Stewardship of the Land I: Scotland's Landscape – the Resource

Stewardship of the Land

IT IS IMPOSSIBLE to write about Scotland and avoid Sir Walter Scott. He probably did not mean the lines below in the way I decided to interpret them, but put it beautifully anyway:[1]

> Land of brown heath and shaggy wood,
> Land of the mountain and the flood,
> Land of my sires!

Most of Scotland's land is wild and remote but still owned by sires, in the sense of 'persons of rank and/or wealth' (rather than 'forefathers'). Many passionate and well-informed works have been written about the Scottish structure of land-ownership, which is feudal in some respects but includes substantial state holdings. This has consequences for priorities when it comes to the use, management and development of land both in the cities and the countryside. The issue of land reform has been spooking Scottish politics since the Clearances and before, and, in spite of a draft Bill on the books, remains a source of concern.

Although most of the population lives in built-up areas, concentrated into the Central Belt round the Edinburgh–Glasgow axis, only a small proportion – about 2 per cent – of the land is actually urban territory. Out there in Scotland's countryside, the stakes in terms of sheer acreage are high. As the demands for changing the ancient rules governing land ownership in Scotland are being transformed into the cautious, legal language of the

Land Reform Bill, urban land has just about vanished from the agenda.

The idea of 'stewardship of the land' is close to the hearts of landowners – all claim to be stewards, farmers on plots of hill land, wealthy owners of huge sporting estates, statutory agencies and voluntary bodies in charge of land. It is attractive for precisely the same reasons that have led to the popularity of phrases about partnerships and sustainability. Stewardship can mean anything and everything, but currently it is usually understood in terms of the new 'agri-environmental' schemes. These will subsidise farmers, but it is not clear that they will be specific about methods and countryside job creation. Without careful management of these schemes, farmers could end up being paid both for conventional, high-yield farming and isolated bits of nature conservation to ensure more income still from tourism.

Stewardship is often taken to mean 'looking after an identified landscape feature'. Sometimes there is a management agreement drawn up with a prescribing state or voluntary organisation, as for instance is the case with Sites of Special Scientific Interest (SSSIs) and wildlife habitats. It is important to involve as many people as possible with fewer formal initiatives – village ponds, woodland, meadows and hedgerows – but all such schemes can have elements of tokenism. It is the kind of 'stewardship' that is very popular as part of development plans: plantings on road and motorway verges come to mind. It is nice to hear the facts and figures about wildlife corridors and thousands of trees planted, but such things do little to mitigate all that tarmac. But perhaps the most common special pleading is that of landowners who argue that simply by running a rural enterprise, be it in farming or forestry or game-keeping, they are by definition 'stewards' of the land. As always, it is how you do it that matters.

Some aspects of the tug-of-war between nature and the survival of a living countryside in an urbanised country demanding 'recreational space' will be dealt with later. Before trying to get a grip on what nature conservation and land reform might mean for Scotland's landscapes, I will look at the intentions and impacts of the organisations responsible for two nature-dependent but not necessarily nature-friendly activities: tourism and nature conservation.

Scottish Tourism and 'Eco-tourism'

A recent report of the European Committee of the Scottish Parliament included a brief reference to 'eco-tourism' among the top half-a-dozen items (Energy Policy, Biodiversity, Waste Management etc.) in the European Commission's *6th Environmental Action Programme.*[2] The European progress assessment – 'Europe's Environment: What Directions for the Future?' – was not particularly optimistic and nor was the committee, although it made the usual reassuring noises. On eco-tourism, it said cautiously, clearly not quite grasping what it was all about: '[Eco-tourism] is an area of particular importance to Scotland given its magnificent environment and cultural heritage. It is estimated that wildlife tourism could generate £90 million ($135m) and employ 4,000 people in Scotland within three years. However, those in the hospitality industry must be made aware.' At the time of writing, some six months after the quoted pronouncement, my inquiry about the Scottish version of an Environmental Action Programme got the following response: '[I am] sorry to inform you that . . . events have somewhat overtaken the Committee. This project was terminated before it began, and was replaced by two major inquiries into reform of the Common Fisheries Policy and Governance of the EU. That is not to say we will not return to the subject matter later, but probably not until the Autumn of 2001.'

The Directives (the EU term for 'legally binding regulations') from Brussels on environmental management have on the whole brought the UK nothing but good, but may have bolstered the wide-spread British conviction that environmentalism is something foreigners do and involves a fearsome bureaucracy. As for eco-tourism, the response of the Scottish Parliamentary Committee suggests that at least in the short term, it is not the right forum. As so often, the emphasis was on the environment somehow 'paying its way'. People often feel this about tourism, but the risks inherent in 'outdoors' tourism can be great, as stories such as the one told below about whale and dolphin watching demonstrate.

All tourism should be eco-tourism. As the Swedish Eco-tourism Society puts it: '[Eco-tourism] treats with care the sources of natural and cultural experiences that attract the tourist, contributes both to the protection of nature and to the economy of local people, is almost always on a small scale, and expects both travellers and travel agents to meet many obligations, all aimed at enhancing the travellers' understanding of issues related to conservation of nature, culture, environment and also to potential local development.'[3]

The quote contains many arrestingly non-Scottish (non-British) emphases, for instance on the word 'obligation'. The main burden of the message is conservation and education, not enjoyment. Dour and moralistic? There was more of same, notably the society's 'Ten Commandments of Eco-tourism', which do not pull any punches:

1. The first concern must be the carrying capacity of the local ecological and social systems.
2. All travel agents should have at least one member of staff responsible for the environment and for the detailed environmental plan for each activity.
3. Local subcontractors must be bound by the commitments to environmental management.
4. Use only accommodation providers adapted to the demands of the environment.
5. Well-informed and knowledgeable guides are crucial.
6. Support the local economy by buying in goods and services in the area.
7. Influence the travellers to adopt respectful attitudes ('Do as the Romans do. . .').
8. Do not buy up people's livelihoods, e.g. objects made from threatened animals or plants.
9. Eco-tourists must be well informed (supply

travellers with details of useful reading and equipment, local customs and practices, the agency's environmental programme).

10. Eco-tourism must support the protection of nature and local development.

Unusually detailed and uncompromising maybe, but it is important to realise that this is not just the outcome of Swedish orgies of group-moralising. By 1994, the realisation that some things about conventional tourism were terribly wrong led to the setting-up of the US-based Campaign for Environmentally Responsible Tourism (CERT).[4] Sure enough, the organisation CERTify touring companies as 'environmentally aware' and hands out environmental awards on grounds similar to the Swedish criteria.

CERT remind us repeatedly of the scale of the tourist industry: 'It is estimated that the world will have over one billion annual travellers by the year 2010.' They go on to state how, in different ways, mass-tourism tends to destroy the very things the travellers want to enjoy and admire. It is hard to imagine exactly what one billion travellers can do in a year. The mind stalls, as the imagination begins to sketch the sheer scale of the enterprise, from aircraft emissions into the fragile stratosphere to water usage and pollution by hotels and the disruption of local ways of doing things.

The voice of Visitscotland – formerly the Scottish Tourist Board (STB) – reaches you from another world. Together with its regional boards, it is the quango in charge of 'the largest business in Scotland', which has annual takings in the order of £2.5 billion from some 12 million visitors and employs about 8 per cent of the population.[5] Over the last decade or so, business has been relatively poorer than before. Visitor numbers have been declining year by year as has income, even though the numbers of relatively affluent travellers is falling less steeply. Again and again, the political establishment and the industry's press release machinery have analysed, exhorted and reformed. In November 2000, an agitated Enterprise Minister sacked the

6. *Ascending Beinn a'Bhuird on the Mar estate, Deeside.*
© Colin McPherson

Chief Executive of STB (as was) and his Chairman followed with a resignation soon afterwards. Six months later, the top jobs had been filled and followed by another round of recriminations and sackings. Recent crises involving the climate (floods) and livestock farming (disease) have not helped. Meanwhile the political masters of Visitscotland have been going to great lengths sell Scotland. Intriguing as it is to watch official panic on this scale at close range, it is also depressing. Any hint of eco-tourism guidance to Scotland vanished, presumably for fear that that anything less than complete freedom to enjoy would be off-putting.

Searching the redesigned Visitscotland website, I found that 'eco-tourism' and 'green tourism' (the title of an STB publication) jammed the search engine, but 'outdoors' and 'wildlife' produced some hits.[6] The results were bleak (mostly just addresses), especially in the context of the language of Scottish tourism, which is awash with romantic vocabulary. One or two of the firms with web-sites of their own managed to be both florid and unhelpful. One of the best sites advertised Shetland Wildlife Holidays. Even so, their catalogue said only the minimum about eco-tourist responsibilities and tended to leave education out of the picture.[7] Could it have something to do with the fact that all the visitors seemed to be of retirement age? But although carping is always worth doing, it would be unfair not also to praise this ambitious attempt carefully to introduce small groups of people to a lovely part of Scotland.

Wildlife Tourism: Whale and Dolphin Watching

When wildlife tourism takes on the characteristics of bear-baiting, one despairs for the future. Some 'safari' excursions come to mind and so, sadly, does whale and dolphin watching off the Scottish coast. This is bad enough, as anyone who respects these remarkably intelligent, inscrutable animals would agree, but in addition I feel that this particular muddle is a generalisable example of the weaknesses inherent in the present voluntary, partnership-dependent, non-directive policies.

Over the years, there have been alarm calls in the press, not only about the Scottish cetacean populations, but also about other colonies along the British coastline and, indeed, worldwide. Whale watching is relatively new – it has grown fast since the mid-1980s – but big business, as a useful report published by the Whale and Dolphin Conservation Society (WDCS) shows.[8] It has been particularly successful in some countries, for instance Argentina, Brazil and New Zealand. These are also among the relatively few that have taken the trouble to pass protective legislation to deal with disturbance and harassment of the animals. By comparison, the UK is a small player: in 1995, there were nine UK sites against Norway's four, although Norway still earned five times as much in revenue. However, it is an interesting and growing business. The Scottish Tourism and Environment Forum (STEF) is keen and sounds admirably concerned about the welfare of the animals: 'Watching dolphins from boats is exciting. We're less sure the dolphins find boat watching so entertaining!'[9]

The STEF *Dolphin Space Project* text properly refers to marine mammal research carried out at the Lighthouse Field Station, which belongs to the Zoology Department at Aberdeen University. Their seven-year-long programme of observing the Moray Firth population of bottlenose dolphins (a rare northern species) has become increasingly closely involved with conservation issues and is linked to and in part funded by the World Conservation Union (IUCN), the WDCS and the EU. The hope is that the Moray Firth will be a European Protected Habitat by the end of 2001. Just in time, it would seem. Although the Lighthouse research publications conform to every rule of scientific discretion, their overall message is that the bottlenose dolphins are under threat, notably from pollution. Their numbers are declining due to various ailments, including the skin sores that afflict all individuals, and new (to the species) self-destructive behaviours such as infanticide.[10]

It has, however, taken journalists to tell us about the full extent of the stress the animals endure, as in Anthony Browne's 'Last dolphins drowning in man's filth'.[11] The burden of Browne's story was not just the continuing flow of pollution into the firth and surrounding sea, nor the unchecked practice of illegal fishing (mainly for salmon) with mono-filament gil nets which harm all marine creatures, but also the harassment of the animals by assorted holiday-makers. The dolphins have been chased by jet-skiers, herded by groups of powerboat owners and pestered by boatloads of tourists, who throw food and other material in their direction. The numbers alone are cause for concern: one tour operator takes 1,500 people per month to see the dolphins and there are fifteen other companies, allegedly pulling in £15 million per year between them.

The picture of crass commercialism is complete and far, far away from the fantasy presented by the ex-STB chairman: 'We are also rewarding the best in environmental and cultural tourism – a chance to look at what attracts people to Scotland from the rich landscape to the history and culture.'[12]

This quote comes from Lord Gordon of Strathblane's Introduction to the winners in the annual competition for Scottish tourist establishments for the eighteen STB Thistle-group awards. Of the eighteen awards, only one group had a green tinge. The three winners were: an estate garden where the tea-room had solar panels on the roof, a guest house with eco-furniture (?) and the Hebridean Whale & Dolphin Trust for its marine discovery centre 'with retail outlet'. Good stuff, as were the ideas awarded in the 'cultural tourism' category (two folk music events and an archaeological museum), but somehow neither convincing nor sufficient. Especially since all the rest was about food, whisky, marketing and e-commerce.

The Media and Selling Eco-tourism

Like so many other organisations, STB have become hypnotised by the role of e-commerce. It is undeniably important. But most travellers search the web for hard data (times, addresses and so on) and make their initial choices by reading newspapers and magazines, or maybe by watching places in films and on TV. I wondered what the Scots were told about eco-tourism, if anything, and looked at a handful of Scottish quality papers (a scan through the tabloids convinced me not to bother).

It was rare to find one of the travel supplements devoted to eco-travel, but it does happen. *The Herald* published one with what seemed like an ambiguous brief that demanded a steady flow of travel descriptions with only rare eddies of fighting talk.[13] Occasionally the tone became positively apologetic: 'Hugging trees is not essential.' An interview with the then Scottish Minister for Transport and the Environment was bafflingly beside the point. 'I never leave home without a set of house keys,' began the minister helpfully and went on in the same vein. The eco-tourist goals were interesting and varied, ranging from Inuit territory to the Scottish Seabird Centre, and there was a jolly, centre-fold guide to the ideas behind eco-tourism. *The Observer* in Scotland ran a shorter version, carrying only one proper article (about Kenya), apart from a patchy review of the problems of world-wide tourism by the old eco-frontman David Bellamy.[14] The rest was 'non-eco', in spite of the front page assurances that this was an issue about 'ethical tourism'.

However, it seems that editors, even in quality papers, feel that not to announce the 'moral' angle is the better policy. In another issue of *The Observer* travel magazine, the Eden project in Cornwall was the attention-grabbing front cover.[15] A well-written article by John Mulholland praised this truly exciting, imaginative project just as fulsomely as it deserves, with a smart sideswipe at the 'dozy countryside'-style provision of main road chain-outlets for tourists. There were a couple of would-be ethical (if slightly risible) tourism stories hidden in the same issue, including an eccentric outing to Cuba of organic gardeners ('Dig for Fidel'). The *Sunday Herald* skiing issue of its colour magazine scored on two counts: it focused on local holiday opportunities – less about Colorado, more about the Nevis Range – and the usual travel hyperbole was tempered by Rob Edwards's sombre article about climatic change.[16]

But this kind of thing is disturbingly thin on the ground, and practically non-existent in other media, unless one counts – as one well might – the ubiquitous wildlife programmes. As already observed, wildlife tourism has built-in ecological trip-wires. Without the safaris, most of Africa's and Asia's nature conservation programmes would probably have gone bankrupt. But the exploitation of local resources of people, land and water is in many cases so gross as to make the whole exercise counter-productive.

Away from the occasional earnest efforts in the quality papers and documentary (rather than travel) features on television, the bulk of commercial tourism is presented with the relentless cheerfulness that only a steady supply of generous freebies can generate in media folk. So far, travel writers clearly do not want to associate themselves with the loony fringe and point out that the most crucial things that could be done to reduce the ecological impact of tourism is to tip the balance away from planes and cars in favour of boats, trains and horses (or camels or donkeys or . . .) and imposing ecological standards on all hotels.

'Managing Visitors'

The Scottish Tourism & Environment Forum (STEF) is dedicated to considering policy and bringing about environmentally sound change. The debate tends to sound hesitant, even rueful. The STEF is a partnership of quangos: Scottish Tourist Board (STB), Scottish Natural Heritage (SNH), Scottish Enterprise (SE) and Highlands & Islands Enterprise (HIE). The STB's special contribution is a Green Tourism Business Award Scheme 'run by Quality Assurance' and not to be confused (I think?) with the environmental type of Thistle Award (see above). I found the general information available in press releases and web sites so opaque it was impossible to understand what the rules were or why particular hoteliers had been selected for an award. However, having sent for the application forms, I found the criteria turned out to be quite tough, bureaucratic in tone and with enough categories and sub-sections to make the keenest manager despair. The environmental standards were thoughtful and wide-ranging, but closer examination revealed that you did not have to be that good. Decent waste management, loft-insulation and up-to-date central heating thermostats, a nice garden, use of local produce and products: do some, or most of that, and your establishment would have a fair chance of a prize, especially given that there seem to have been only twenty entrants in 1999.

This is at least a start, but the promotional material should also target the tourists. It is surely possible to do more to outlaw behaviour likely to harm 'environmental order and safety' in the same way as holiday makers in cities have learnt to conform as a matter of course to all the rules for built-up areas – they cross by the crossings, avoid having noisy picnics in museums and so on. 'Nature' should not be seen as a no-rules, free-for-all playground.

The idea of setting up Tourist Management Programmes (TMPs) sounded like a way to bring about joint responsibility.[17] Each TMP is run by partnership consisting of the regional TBs (Tourist Boards) and LECs (Local Enterprise Companies), and is joined by any other regional or local partners prepared to take an interest. TMPs are about sustainable tourism: '[Reductions in] traffic congestion, erosion and litter' and the enabling of new measures by 'Dialogue . . . pooling resources and expertise . . . avoiding duplication . . .' It sounds very well intentioned, but the current lot of seven TMPs and two TMP-related projects leave one with the overall impression of caution and slow progress – very slow in some cases. The initiative seems to have got off the ground in 1992–3 and abbreviated lists of the successes for each area are included in the Notes.[18] The main web-site does not carry a 'last updated' indication, but does not give any details beyond 1998. By then some areas had got nowhere in particular ('better networking', 'describing habitats'), while others had been busy with tidying up, creating paths and street fairs and so on; and two places, Callanish and Trossachs, had initiated bigger projects (Discovery/Visitor Centres, bus services).

The Scottish tourism business is notoriously in deep trouble these days. It is not short on willingness but seems woefully short on efficiency and imagination. To find out more about how eco-tourism is faring in this climate, I went to ask an expert.

Interview with Rory MacLellan

(senior lecturer in Tourism Studies at the University of Strathclyde, Glasgow and an expert on eco-tourism)

I was grateful for the chance to speak to Rory MacLellan, who is constructively critical of Scottish tourism from an environmental perspective.

The interview started with a discussion of the appearance in the early 1990s of references to [environmentally] sustainable tourism in the business of the Scottish Tourist Board. Rory MacLellan referred to an umbrella organisation called the Scottish Tourism Co-ordinating Group (STCG) and spoke with insider assurance about its original TMP objectives. He is currently a member of the Tourism and Environment Forum (TEF; previously the Tourism and the Environment Task Force). He said that environmental tourism in general was not well understood at the time, and the role of the TMP programmes was unclear. They were dependent on individual commitment in the localities. The new initiatives should have been monitored but never got much beyond base-line studies by independent consultants.

MacLellan has described and discussed these studies in several publications.[19] On the whole, he feels that the studies were done on the cheap, for instance based on feel-good questionnaires and one-off counts of visitor numbers. Proper process auditing similar to an Environmental Impact Assessment would have been a complicated and time-consuming affair and hence expensive. The inevitable consequence was that this did not happen anywhere. But without investigation there was no guidance about effective ideas. For instance, there were ten partners in the Stirling-based team that managed the TMP in the National Park-rated Trossachs, but the contributors to the evaluation of visitor impact on the physical environment did not include one ecologist or naturalist, not even an amateur naturalist.

The management of Scotland's tourist assets lacks positive leadership and directed funding, Rory MacLellan feels. This in turn hampers the development of 'true' eco-tourism projects. Ineffective monitoring is a medium term problem, but the most serious immediate one is the way money tends to circulate between lead agencies, who anyway hold the purse-strings. Typically, the partnerships are dominated by LECs, Area Tourist Boards (ATBs) and regional SNH branches, with the local authority presiding. The larger wildlife charities, like RSPB and WWF (Scotland), can be influential when their interests are concerned. Of the agencies, only SNH has nature conservation at the top of the agenda. The charity representatives are important, because their single-mindedness can compensate for the influence of local politicians, whose priorities are jobs, area prosperity and voter-friendly compromises.

MacLellan is concerned about the fragmentation of decision-making and the lack of common strategic goals. He feels that this can be the case at many different levels, from local groups to the Scottish Parliament. The insecurity in the system is neatly shown in the TMPs: he argues that anxiety is a check on progress. The agency and charity representatives are not linked so much to the locality as to their own hierarchy – 'What have you achieved?' hangs over them from the start, asked by their bosses rather than by local people. Interested locals are at best on one- or two-year part-time contracts to working parties and implementation committees and are often giving freely of their own time. Occasionally, a project works well. MacLellan mentioned in particular the Skye project (restoration of a Skye–Lochalsh footpath) and the support for local events in St Andrews.

We went on to a large-scale management project, the Cairngorm Summit TMP, and a related funicular railway plan discussed again in the next chapter. He felt that the Cairngorm developments have become counterproductive, both in human and environmental terms. The 'railway and café' idea was first put to the TMP Committee by the Badenoch and Strathspey LEC, as a part of visitor management. After almost three years (1995–8) of heated debate, it was adopted at a meeting in the Sports Council's Glenmore Lodge, the original TMP became dormant and SNH had to insist on strict control of access to the summit. MacLellan sympathised with SNH, but felt that if day-trippers are to be ferried up in the first place, the boundary round the café is too tightly drawn. The rows have not stopped since and have led to notable resignations – one of the most significant was that of the ecologist Alan Watson, who has spent a lifetime monitoring the Cairngorm habitats – and recurring allegations of corrupt decision-making. The final comment on this saga was that the main weakness of the 'partnership' was the lack of precisely defined and enforceable conflict resolution mechanisms.

MacLellan commented on the remarkable fact that John Muir's homeland was the last in Europe to accept the need for large nature conservation areas of the National Park type. He identified three main reasons: land ownership patterns, the notion that 'Scotland has enough wilderness not to need Parks' and concern about the socio-economic deprivation in areas distant from the Central Belt. The National Park proposals from 1990 were turned down by the Conservative administration for reasons related to all three factors. The current legislation is partly due to a political change of heart, and partly to pressure by voluntary organisations in the wildlife and conservation sector. Even so, designating the first two National Parks in Scotland was a unilateral decision by the then First Minister without any prior consultation with agencies such as SNH. An opportunity was missed, both with regard to the wilderness, biodiversity and to sensitive tourism, and now the two main Park areas are degraded ecologically and lack proper infrastructure for visitors.

Nature – the landscape – is a resource in the context of tourism. It is also a resource for outdoor sports and leisure activities. The step between tourism and sport is a short one, and the interests of the two sectors are closely linked. They share a need for planning and resource management, as well as for effective and targeted promotion. There are elements of high principle in both, but even more so in the case of sport. Does it have room for long term perspectives and duty of care for the environment?

Sport as Morality and Machine

On one hand, sport can be about 'countryside recreation' and offer an urban population opportunities to learn what nature is really like. From that point of view at least, it is important to many that the experience should be reasonably immediate and 'real'. 'Unreal' countryside sport is the kind subject to the ministrations of guides and Visitor Centres, and bolstered by plenty of 'facilities', like surfaced tracks and cafeterias. A sizeable minority, however, wants to feel able to endure and cope with challenges different from those of the 'built environment', from cycling against the wind to ice-climbing towards a distant summit. Going to stay in the country for a holiday is often driven by the same kind of motives, including a wish to be closer to something almost lost. For many, the loss can be made up for in green belt and 'edge' areas round town and cities where parks, playing fields and path systems along canals or disused railway tracks are sources of new feelings and sensations.

On the other hand, sport is a consumer activity and big business, especially the job-creating, money-swilling world of professional sport. In the context of tourism, this too can affect countryside sport and recreation. In the microcosm of sport, a familiar pattern is repeated: the state has largely left the money-making side to regulate itself and the rest has been bundled into a ministry with unclear responsibilities, which until recently included culture and environment. The tasks of advising on strategy, reporting of the current state of affairs, managing grants-in-aid and generally keeping track of the ruling bodies of individual sports have all been handed on to the Scottish Sports Council or, as it is now called, SportScotland.

As the new millennium was drawing closer, SportScotland, together with a substantial strategy advisory group, produced a strategy document with the challenging title *Sport 21: Nothing Left to Chance.*[20] It is an important document for many reasons and reading it with an environmentalist's bias had its own interest.

The *Sport 21* consultation exercise seems to have been comparable to others carried out in the world of quangos. Consulting is part of received wisdom and the exercises have become expensive and elaborate. Of course, they are only useful in proportion to how well they are conducted and analysed. I have not examined the process of the *Sport 21* consultation, but it looks typical: eclectic questions handed out to everybody, from fellow quangos and specialists (e.g. governing bodies of various sports) to less obvious consultees (e.g. Action on Smoking and Health). The final document contains no systematic review and no statistical information on who thought what about what.

Looking for 'the environment' and 'environmental sports', the headline 'Creating a New Sporting Environment' seemed to announce exactly what I was looking for. It turned out to be about the need a 'national framework for sport', prioritisation of existing resources (no, the resources did not include the landscape) and better co-ordination of effort. 'The Four Key

Challenges for 2000' were: a National Physical Activity Taskforce, a Scottish Institute of Sport, something to the effect that governing bodies of sport should be given a lead and most hopefully from my point of view, 'Local Planning for Sport'.[21]

The selling off of sports grounds has been a scandal with a UK dimension. Scotland has seen marked erosion of community sports facilities, as a net result of patchy planning – excellent in some bits and terrible in other. The section on local planning dealt mainly with the need for facilities, often of the kind that is lit, heated and/or covered in impenetrable artificial turf. There was no mention of the energy usage of such places, or the waste of materials. This is all the more remarkable, because there are sports buildings, in Scotland and abroad, which are both good-looking and environmentally scrupulous. Some of the Scottish examples have had the support of SportScotland, such as the Gaia Group's interestingly constructed McLaren multi-use hall in Callander.[22] SportScotland has a creditable record of informing itself about ecological building techniques.[23]

7. The McLaren Community Leisure Centre in Callander, completed in 1999. Constructed following ecological principles, this is the largest building in the world using dynamic insulation (pore ventilation). © The Gaia Group

Still, the strategy document focused on 'facilities', without qualifications. The section on 'Sport and Recreation in the Countryside'[24] takes up two and a half pages out of 150. I was helpfully sent some more booklets, which looked more promising: *Earth, Wind and Water*[25] and a special publication on water sports, *Calmer Waters*.[26] In many respects, *Earth Wind and Water* treats the countryside as a facility. Out of nine Guiding Principles, 'sustainable development of the natural environment' comes in as Principle 4.[27] Generally, SportScotland seems determined to get its way with the land, perhaps most strikingly expressed in Principle 3, which is quoted in the box:

> The Council considers that all of Scotland's countryside, regardless of existing or future designations, should be considered as potential locations for sports within the parameters of these guiding principles.

The two countryside sports publications elaborate the ideas enshrined in the Guiding Principles. 'Access' amounts to an obsession, pursued almost more relentlessly than in the literature of the Ramblers' Association. However, SportScotland is necessarily less focused and perhaps also less articulate than the walkers. *Earth, Wind and Water* speaks in at least reasonable terms about sustainability, but clearly influential views in the organisation have led to statements such as '[SportScotland supports] the overall concept of sustainable use of the natural environment but opposes [its] automatic application.'[28]

Still, SportScotland believes that no place, whatever designation for nature conservation it might have, is unsuitable for sport. It speaks of 'sites' in terms of development, that is 'environmental site capacity' (a dangerous concept in the wrong hands) and also of 'management techniques', 'regional strategies . . . for noisy sports' and 'access to air space [and] new airfields with minimum restrictions on aircraft types'. There are demands for special facilities: skiing, subaqua (car parks), hang-gliding (landing sites), target shooting and so on.[29]

Calmer Waters is more precise, more helpful and generally less demanding in tone. The 'sports first'-drive occasionally comes to the fore, mostly in the insistence on effective 'management strategies' rather than disallowing sporting developments. Generally, with its moderate tone and helpful illustrations (plain maps) with examples of thoughtful existing developments, it presents a more hopeful picture of what could be done. The all too neglected practical management of recreation in nature must take account both of where people live in Scotland (in cities) and the need to have some actual nature left that is not tracked, paved, and turned into 'facilities'.

SportScotland has sorted and listed 'sports' into Levels 1, 2 and 3.[25] It is obvious that Level 1 includes prestigious and/or wealthy sports like athletics, football, rugby and golf. Cricket, curling and bowling are in favour too, though the reasons are less clear, given that the also-rans include popular folk sports like rowing, gymnastics/fitness and skiing (Level 2) and mountaineering and wrestling (Level 3). Why is badminton in Level 1, but tennis in Level 2? What is health & beauty (Level 3) doing there at all? These rankings stayed inscrutable, even after talking to SportScotland, but it looked as though most low-tech countryside-based and family sports are out of favour. Countryside exceptions are golf and possibly swimming, though it is not clear if that refers to family-friendly pool facilities in- or outdoors.

The language of sport is curiously old-fashioned. On one hand, it is full of the high morality of endeavour and achievement and on the other, of the commercial virtues of getting paying bums on seats. The scenic and profitable sport of golf is a good example.

The Story of Golf

Golf is an endearingly difficult game. People who can play it at all well, love it – as my arthritic friend Bill said after another couple of painful hours with much outward rotation at the hip joint: 'They'll have to cut my legs off before I stop playing golf.' It can be played by all age groups and all income groups, for enjoyment of it depends only on having a couple of different clubs and a supply of small balls – and a stretch of rolling but essentially smooth land. Like all sports, golf had its time of innocence, when this is exactly what it was like. The first great Scottish links courses were nothing but the hinterland of sea-side dune landscapes and the biodegradable kit beautifully but relatively inexpensively made from wood and leather.

The growth of this popular pastime into one of the world's wealthiest and most prestigious sports has been remarkable. However, its environmental impact has been little short of disastrous, ranging from waste of material resources used for all the equipment needed even for ordinary weekend golfers, to demands for clubhouses, car parks, changing rooms and other facilities. Golf-course maintenance alone is enough to bring tears to the eyes of nature lovers, because behind the pristine greenness of the grass lie many man-hours of spraying with assorted fertilisers and pesticides, eliminating live nuisances like rabbits and birds and generally manicuring the landscape for exclusively human use.

The environmental aspects of golf are looked after by a complicated partnership called the Scottish Golf Wildlife Initiative (SGWI).[30] Can golf excuse its increasingly pervasive use of land in Scotland by becoming 'environmental'? Maybe not, although after having read some of the literature published by the SGWI it is easier to believe. The SGWI is supported by every conceivably relevant organisation, including SportScotland of course, and a range of others, from SEPA to the Butterfly Conservation Society. It would clearly be very reassuring to have skylarks and fritillary butterflies nesting in the semi-rough, but a scan of the latest news on the web-site of the British and International Greenkeepers Association was not encouraging.[31] The expensive publicity given to turf management – read pesticides, fertilisers, mowers with very sharp, very close-cutting blades and many other kinds of machinery – suggested that plant- and wildlife-friendly golf might be a long way off.

Golf tourism is big business indeed. The golfing fraternity is very keen on the sport, but is also outstandingly wealthy and well travelled compared to most other sport supporters. No other sport has developed such a network of glamorous real estate and tourist facilities, including championships, hotels and retail shopping. The way the Scottish have chosen to promote golf tourism is quite remarkable for a people with deep-rooted – and often endearing – reservations about 'running a service economy'.

But it is not difficult to see the reason why. Large tracts of pointless-looking land, including arable farmland (who needs farms these days?) can be made to pay their way by being turned into sterile grassland. Construction firms like golf developments, because riding on the back of local authority approval of 'recreation' and 'local jobs' comes the opportunity to build the hotels, chalets, country clubs, gymnasiums and parking places that the golfers demand.

The Kingask development (St Andrews Bay) near St Andrews, Scotland's golfing Mecca and already in proud possession of six major golf courses, is a relatively recent case in point.[32] The Scottish media took an interest in what had become a protracted battle between determined people in the community supported with evidence from agencies such as SNH and Historic Scotland, and the developers, backed by the majority of a divided local authority, the regional Enterprise Company and Tourist Board. The criticism of the development, a 'make-believe golfing Brigadoon' built in an already over-supplied region tended to be balanced by recognition that the more an area specialises, the harder its gets to disengage – 'having sold themselves to the world as the Home of Golf, with must-play courses and [St Andrews] as a cute historical backdrop, the town and its people now have to live with the consequences.'[33]

The point is of course that by investing in non-sustainable enterprises, everyone will lose out sooner rather than later. There is nothing local or stable about developments like St Andrews Bay. Expensive 'Big Golf'-facilities are at risk when international markets shift. If the rich people from abroad do not come, the St Andrews Bay Country Golf Club and all its clones will be of no value to the community. This is the real objection to developments of this type, but curiously one that is rarely used. Maybe no one likes to face the unthinkable prospect of the Americans and the Japanese not wanting to come to Scotland, or worse, not being rich any more.

More about SportScotland

I did interview a representative of SportScotland but it was felt that I 'misquoted and misrepresented' the person who had agreed to talk to me. The text below is a summary of the facts, partly based on my notes. Nothing I say can be assumed to represent the views of SportScotland.

SportScotland distributes Sports Ministry money and Lottery largesse to the sports in Scotland on both regional and national bases, and there the ranking of a particular sport matters a great deal. The funding comes in the form of, for instance, project grants or spending allocations to governing bodies.[34] The differences can be very great, that is little or nothing for some sports and substantial backing for others. The criteria used to evaluate applications include the sports rank order system mentioned earlier. The environmental credentials of a project are taken into account. Environmental impact assessments are seen as a matter for the local authority or SEPA, rather than SportScotland. It can however lodge objections to projects even after planning permission has been granted, and has a planning team, the duties of which include keeping a log of planning applications. The most common and most deplored form of local change tends to involve development on sports grounds like tennis courts and school playing fields. Changes in provisions should not entail any overall reduction in facilities and improve provision for local sports in the long term.

SportScotland runs three national training centres, with one in the Highlands (Glenmore Lodge) and two in Ayrshire (both in Largs), but the proportion of the staff responsible for countryside sports is relatively small. The organisation clearly feels that all sports should have access to all suitable areas and competing interests should accommodate, rather than exclude sporting use. The internal debate seems pro-environment, but the sheer pressure on the 'countryside sports'-staff means that their contributions to the many debating fora may be less significant than they would wish.

Some of the partnership fora involving the SportScotland countryside staff have been listed in the Notes.[35] The number of meetings, working groups, seminars and reports is very large. At various times, the organisation have taken decisions in favour of: the funicular railway to the Cairn Gorm summit; jet skiing and other motorised water sports on Loch Lomond and other waterways; various forms of cycle racing on the Roman road in the Borders. Other decisions include: no restrictions for sporting access to farmland and to water, inland or sea; and no restrictions on motorised sports or on access by car. SportScotland has supported guidelines on good management practices, for example putting really intrusive sports such as car and motorbike racing out of harm's way (for instance in disused open-cast mine workings) and recommending the use of LPG[36] instead of petrol. Arguments in favour included: the wear and tear on the Roman road was due to agricultural vehicles, not sporting ones; the Cairn Gorm funicular was a good thing in itself, but the summit access restriction, which was the fault of SNH, annoys everyone; the Loch Lomond water has been analysed by SEPA without any evidence of hydrocarbon pollution; the erosion of its shores is natural and not due to vehicle-generated waves.

Also, the general aggression against motor sports can be class related, as those who enjoy these sports are seen as yobs by middle-class ramblers, who are very insistent on their own rights.

The Official Landscape

Every organisation must have a set of policies and officials must be the organisation's loyal and effective spokespersons. It is up to the public to impose expectations of clarity, accessibility and acceptability of these policies. For instance, the listed arguments in favour of sport cannot be uncritically accepted. If the foundations of the Roman road are shaken by farm traffic, then the answer is to discuss restriction with the farmers and not to open the road – lovely for walking and well, just lovely – to wheeled sporting folk. The Loch Lomond water measurements did not show zero hydrocarbon residues in the water, as described in the next chapter. It is apparently true that the main cause of shore decay is natural erosion, but the answer is not metal net-reinforced piles of rubble and more motorised water traffic, but a total ban on anything that might contribute to the eroding process. As for the funicular railway, it too is discussed elsewhere, including the next chapter and in the Tourism section of this.

Official tourism is failing the Scotland that according to the Visitscotland web-site links still has 'one of the last wildernesses in Europe – through the soaring beauty of Glencoe to the idyllic charm of the isles, and from the crashing waves of the northern coastline to the silence of the windswept moors . . .' (Skye and The Highlands) as well as 'one of Europe's great cultural destinations . . . complemented by the beautiful countryside along the River Clyde from its source, through garden valleys to the sea' (Glasgow and the Clyde Valley).[5] This failure is still on a fairly modest scale, nowhere near the ruthless selling that betrayed, say, Costa Brava or Hawaii. But unless Visitscotland starts believing in its own rhetoric, it will copreside over a process that will cause much of the 'last great wilderness of Europe' and Scotland's 'cultural destinations' to become tatty, worn and vulgarised – and so in the end disappoint everyone. Instead every square metre of tarmac should be costed with its environmental impact in mind, as should every new holiday-village development and every laissez-faire waste collecting arrangement. By the way, the idea that a guest house should be given a prize for recycling glass and cooking with locally sourced raw materials is bizarre. Rather, STB should think of some way to penalise establishments that fail basic tests of that kind.

The brief for those who market Scotland should go beyond selling 'Scotland – the Brand'; it should truly include stewardship of the land. One stewardship policy that is currently being tested is the new National Park legislation and the management of some of Scotland's most beautiful and culturally rich landscapes.

Notes

1. *The Lay of the Last Minstrel* by Walter Scott, 1810 Canto 6th, ii.
2. 6th Environmental Action Programme of the European Commission, Work Programme for March 2000–March 2001. 'Europe's Environment:

What Direction for the Future?' http://europa.eu.int/environment/forum/

3. Ekoturismföreningen (Swedish Eco-tourism Society). www.ekoturism.org.ekoforum/ Translations by Anna Paterson.
4. Why the CERT Initiative? Campaign for Environmentally Responsible Tourism. www.c-e-r-t.org
5. www.staruk.org.uk/ Key site for research on British Tourism, with regional figures up to 1999.
6. Discover the Best of Scotland: Outdoor Holidays in Wild Nature. www.visitscotland.com/outdoor/wildscotland
7. *Shetland Wildlife Holidays 2001* [catalogue].
8. *The World-wide Value and Extent of Whale Watching* by Eric Hoyt, 1995, pp. 1–36.
9. The Dolphin Space Project. Whale and Dophin Watching. Tourism and Environment Forum. www.greentourism.org.uk
10. 'Evidence for Infanticide in Bottlenose Dolphins: an Explanation for Violent Interactions with Harbour Porpoises?' by I. A. P. Patterson et al., 265: pp. 1167–70, 1998; and 'Combining Power Analysis and Population Viability Analysis to Compare Traditional and Precautionary Approaches to the Conservation of Coastal Cetaceans' by P. M. Thomson et al., 2000, 14: pp. 1253–63.
11. 'Last dolphins drowning in man's filth' by Anthony Browne. *The Observer*, 27 August 2000.
12. 'The Power of Scotland' by Lord Gordon of Strathblane. Introduction, The Scottish Thistle Awards, 7 November 2000.
13. 'Whose World Is It Anyway? Decision Time for Every Traveller' by Trevor Grundy, Crain Watson, Julie Davidson et al., *The Herald*, 20 November 1999.
14. 'Escape Guide: Ethical Tourism' by David Bellamy and Jake Grieves-Cook, *The Observer*, 27 June 1999.
15. 'Welcome Dome' by John Mulholland. *Escape Magazine*, *The Observer*, 11 March 2001.
16. 'On a Slippery Slope' by Rob Edwards. 'Chill Out: Snow Sport and Style', *Sunday Herald Magazine*, 25 February 2001.
17. Tourism Management Programmes www.greentourism.org.uk/projects/tmp
18. Tourism Management Programmes have been set up in Nairn, Pitlochry, Bute, the Trossachs, Kintyre, St Andrews and Callanish; and the Related TMR projects are: Skye Footpaths Initiative and Upper Deeside Access Trust. Key successes: parking, 'interpretation' signs, beach access paths, promotional material (Nairn); better communication (Pitlochry); tidying, creating paths and cycle/bridleways, rural development and visitor facilities (Bute); improvements to villages, footpath repair, establishing traffic and visitor monitoring systems, a bus service, a Community Forum established, Discovery Centre, wildlife tourism initiative (Trossachs); signposting, plant and wildlife habitat

descriprions (Kintyre); 'interpretation' signs; beach tidying, a coastal path, floodlighting, floral enhancements, fairs, transportation, surveys (St Andrews); Visitor Centre (Callanish).

19. 'Tourism and Scotland's Environment' by Rory MacLellan, in *Tourism in Scotland*, 1998, pp. 115–34.
20. *Sport 21: Nothing Left to Chance*. Scottish Sports Council, 1998.
21. Ibid., p. 18.
22. 'The McLaren Building: Design Objectives', by Howard Liddell, in *Dynamic Insulation – Past Present and Future*, 2000.
23. *Pore-ventilation: Sports Halls. A Study for the Scottish Sports Council* by Howard Liddell et al., 1995; *Sports Buildings and Sustainability*. Seminar between the Scottish Sports Council, the International Union of Architects and the Royal Incorporation of Architects in Scotland, 1997; *Sports Facility Seminar: Building Lifestyle*. SportScotland, 1999.
24. *Sport 21: Nothing Left to Chance*. Scottish Sports Council 1998, pp. 119–23.
25. *Earth, Wind and Water. Statement of the Scottish Sports Council on Policies for the Planning of Natural Resources for Sports and Physical Recreation*. Scottish Sports Council, 1995.
26. *Calmer Waters. Synopsis of Guidelines for Planning and Managing Watersports on Inland Waters in Scotland*. Scottish Sports Council, 1997.
27. *Earth, Wind and Water. Statement of the Scottish Sports Council on Policies for the Planning of Natural Resources for Sports and Physical Recreation*. Scottish Sports Council, 1995, pp. 4–5.
28. Ibid., p. 14.
29. Ibid., pp. 323–4.
30. Scottish Golf and the Environment: The Initiative. www.scottishgolf.com/environment/Contributing golf and sporting bodies include Sport-Scotland, British Institute of Golf Course Architects (BIGCA), British and International Golf Greenkeepers Association (BIGGA), Scottish Golfcourse Wildlife Group, Scottish Golf, Sports Turf Research Institute (STRI), US Golf Association, Highland Golf Development Group and the Committed to Green Foundation (about 'education, research and practical initiatives concerning ecology, environmental management and sustainability in sport').
31. Greenkeeper International – All the Latest News and Features from the Turfcare Industry. The British and International Golf Greenkeepers' Association. www.bigga.co.uk/greenkeepers
32. The St Andrews Bay hotel, club and golf course development on a 500-acre site of arable land near St Andrews. This is a £50 million investment, bringing irrevocable changes to a quiet rural area.
33. 'The Golf War' by Trevor Royle. *Sunday Herald*, 11 July 1999.
34. Many (sixteen, in 1998) Scottish sports are managed by governing bodies responsible for such aspects as event management, competitive strategies, rules and standards, and, when financially possible, training and development.

35. Apparently hundreds of partners, including: CoSLA; SE & HIE; Forward Scotland; Access Forum; Access Forum for Water; STB; Tourism & Environment Forum; SNH; various National Park Fora/Working Groups; Scottish Outdoor Recreation Network (SORN); Disability Scotland; Health Education Board for Scotland (HEBS); the Scottish Water Authorities; Scottish Cycle Forum, Ramblers Association and all the other associations/organisations for outdoor sports (personal communication).
36. Liquid Petroleum Gas, a low emission fuel currently not widely available.

6

Stewardship of the Land II: Conserving or Marketing?

Scotland's National Parks Policy

THE FACT THAT SCOTLAND is one of the few countries in the world without National Parks is mildly astonishing, but from nature-conservation point of view not necessarily a bad thing. Designating a landscape 'National Park' is just an administrative and management tool with many different interpretations, nationally in UK and internationally. Essentially it is about planning with some degree of compulsion thrown in – 'regulation of land use and development', 'enforceable access codes', 'non-consensus based planning decisions'.

The interests of local people often become the most really difficult factors to complicate nature-first planning. Does living in and/or owning a part the countryside give your needs and views priority? Can the countryside be carved up in an equitable way between the different and often conflicting interests in it? There are no methods a democratic state can use systematically to override the will of the local population, although negotiation, persuasion, financial incentives and – on limited issues – coercive legislation might in various combinations win some battles, if not the war. Because I will come back later to the long Swedish experience of National Parks, here is an instructive example of a self-confessed failure. A zoned plan had been prepared in 1988 to create a vast National Park in the far north Kiruna municipality. 'Zoning' is a device for minimising conflicts between economic, recreational and conservationist goals. After six years of negotiations with the local people, the whole plan was scrapped. Officials from Swedish Environmental

Protection Agency wrote a candid summary of the reasons and included a helpful list of what was meant by 'zoning' in this case:[1]

> The zones were: wilderness zone, no buildings or tracks. Trail zone: tourist cabins and trails for hikers and snowmobiles. Preservation zones: research only in strict reserves closed to the public. Snowmobile-restricted area. Highway zone: within 3 kilometres of the main road with hotels and pistes. Lake Torneträsk: special water use regulations.
> The proposals included a comprehensive package of nature protection, eco-tourism and local benefits, and yet the idea was sabotaged by the people living in the area. They wanted the mountains to themselves, with no restrictions on their outdoor recreations, snowmobiling, fishing and hunting. Now many conservationists are wondering whether local interests should be taken into account to such an extent that even a National Park plan can miscarry.

Kiruna has still got mines, commercial railways and an active local population, but it is all subject to the same reductive influences that are depopulating the Scottish countryside: urbanisation, unprofitability of major industries and poor communications.

One possible way to save existing rural landscapes is simply to wait for market forces to empty the countryside of everybody who needs land to make a living. Instead the crofts, farms and stately homes will be bought up by people whose city life-line is long and flexible enough – retired folk with decent pensions, people whose work can be done over telephone lines, arts and crafts workers and, of course, those who manage nature for visitors, from countryside rangers to hoteliers. So, we might get our nature, but at a price. Scotland is a small place, and even the biggest nature reserves in the world cannot sustain the systematic pressure of tourism. By exchanging difficult, possessive and determined landowners for more compliant inhabitants, we will enter into a new relationship with the land.

It seems likely that without an explicit rural development

policy, there is a real risk that the landscapes will turn into managed showpieces, rather like the postcards and calendar views. There are two main goals: to avoid further degradation of the environment and, much harder, to find profitable but environmentally acceptable things for people to do. The idea that people should be manipulated into staying put in the countryside is not a new one. However, the mixture of job-creation schemes and financial incentives (of which the Common Agricultural Policy is probably the starkest failure) have not always been effective. These issues will turn up again in the sections on land reform and 'environmental economics'.

Countryside management by zoning is closely related to the question of survival for a genuinely land-dependent population. If people are to get a living from the countryside, visitors will not always be welcome. Even worse, I suspect, will be the conflicts between groups of visitors with different interests. The Swedish example hints at some of these, for example tourists in hotels and on pistes versus researchers on closed-off estates, and the problems are neatly embodied in the snowmobile row. Most local people in that part of the world have got one, even young children, and they all swear that living where they do would be unmanageable without this practical, tough mode of transport. Some tourists enjoy them because they are fast and fun and exciting to ride. But many others, locals and visitors alike, truly hate snowmobiles for being destructive to the land, terrifying for animals, seriously polluting and unbearably noisy. What to do – but zone, license and cajole.

Zoning can only work if we assign a value, legal and financial, on the countryside. In a sense, the equivalent already exists in built-up areas. The main difference between town and country is, I believe, that the city has got mainly costed resources, even when free of overt charges, while most things in nature are still unpriced commodities.

How to Choose a National Park

Built conservation areas are usually well policed by fond, eagle-eyed guardians. Urban planning regulations often seem more detailed than those applying to the countryside, but everywhere, planning stringency tends to be in inverse proportion to the size of the project. I have not been able to find a study comparing the effectiveness of what is after all known in Britain as 'town and country' planning. The term has a long tradition but is more valid now than ever as most modern planners have come to recognise that: 'the way we fashion the urban (built) environment also affects the rural and natural environment . . . we should perhaps rename our activity "environmental planning".'[2] This is an important statement, although perhaps easier to say than to make operational. For 'natural' read 'rural': wilderness, with its romantic and 'not-human' connotations, is exceptional in Britain and even in the Scottish Highlands. Maybe this absence of the wild contributes to a utilitarian attitude to the great outside.

The previous history of the National Park idea in Scotland was briefly referred to by Rory MacLellan, and will be discussed again in this chapter by Roger Crofts and Christopher Smout, who has set the Park idea into a historical perspective in his book *Nature Contested*.[3] The National Park models vary but agree on some things: the designation should confer a high order of protection of natural and cultural assets; conservation should come first in conflicts about development (the 'Sandford principle' in England and Wales); there should be an independent Park Authority with planning powers; local people should be prepared to forego some of their rights in exchange for a measure of influence over the Park Authority and the advantages that protected area status would bring. Of these four, the first three criteria have been close to the hearts of conservationists for many, many years and have been rehearsed in the *Consensus Statement on National Parks* by the group of NGOs called Scottish Environment LINK.[4]

The Park idea seems a reasonable solution to the problem of how to control nation-wide demands for development land, but the protection has not been comprehensive. T. C. Smout has listed some of the many instances where 'unnatural' developments have encroached on the English Parks.[5] Scottish conservationists are concerned that the present legislation –

The National Parks (Scotland) Act 2000 – is going to provide weaker protection than in England, but local people, local authorities and some powerful quangos worry that the Park designation will prove too rigid a constraint on their plans. Maybe there will turn out to be a 'Third Way'. The Park designation might make people think again about how to create an economic eco-system that includes the local landscape.

Two areas have been selected as the first potential Parks: Loch Lomond on the edge of the Central Belt, with its lovely wood and moorland setting of the Trossachs Hills, and the Cairngorms area, a group of wild high mountains extending into rolling uplands. Both are remarkably beautiful landscapes but the risks they run and their vulnerability are different.

Picture Postcard: Loch Lomond and the Trossachs

The landscape of Loch Lomond and the Trossachs centres on the great loch. Around it lies partly farmed, partly afforested upland, rising to high hills such as Ben Lomond and Ben Vorlich. The area is a much loved holiday destination for the populous Central Belt and in particular for the people of metropolitan Glasgow. It has been and is a national icon, but paradoxically, designating it the first National Park might be the beginning of the end for its period of relative peace.

With the primary legislation in place and accepted with guarded enthusiasm, the partnerships led by SNH went on to explore the specific conditions that should apply to the two first Parks: Loch Lomond and the Trossachs (LLT) first, then – hopefully – the Cairngorms (Cg). The objections to the National Parks Act focused on the assumption of compatibility between the four elements of 'sustainability': local economic development, social equity, healthy recreation for all comers and nature conservation. Conservationists argued that the legislation favoured local people and economic development, and was correspondingly weak on the national interest in protecting nature.[6] The possibility of 'zoning' was not clearly set out, at least not in the LLT consultation document.[7]

The document stresses the balancing act needed to make

sustainability work, but factual information is a bit sparse. The three maps show 'Existing designations and features of interest', 'Potential areas [for inclusion] for the National Park' and 'Potential division into sub-areas', but still fail to help with the 'Key issues for consultation'. For instance, the document maps do not show the River Leven and its barrage, a crucial feature in nature conservation on the loch that will be discussed below. The consultees are meant to give special thought to the Key Issues, for instance 'Should Flanders Moss and the agricultural land round Loch Ruskie be excluded?', but neither the loch nor the moss is shown on any map.[8] When I followed up this Key Issue, picked for no special reason, it turned out that Loch Ruskie and its agricultural land gets no area mention at all, but Flanders Moss is rated a Primary Consideration area (i.e. rated between the 'Core' and 'Secondary Consideration' areas).[9] The moss is a National Nature Reserve (NNR), with several Sites of Special Scientific Interest (SSSIs), having as well the international designation Special Area of Conservation (SAC). There is no mention of the attempts to save the moss from becoming dried-out scrubland by intensive agricultural drainage or the sheer rarity of the site. The moss is under the protection of the Scottish Wildlife Trust's Peatland Conservation Campaign, initiated because over 90 per cent of peatlands have been drained, planted or dug up to be spread on gardens.[10] Maybe only those who know these Key Issues well are meant to state their opinion, but the consultation exercise seems more inclusive than that. Defining sub-areas where specific activities can be encouraged or discouraged is zoning by another name. It could be a useful alternative to the troublesome deals brokered between the four local authorities, whose boundaries meet in the area.[11]

The loch is described as offering 'opportunities for water-based recreation of national importance'. This could have been used as an example of the kind of difficulty zoning might help to solve. The problem is sporting use of motor boats and other motorised recreational water vehicles such as jet skis. These alarmingly fast, noisy and toxic means of having a good time mean bad times for many others, from people to plants and water creatures. A recent study (why not before?) carried out by the ecologists at the Research Field Station on the edge of the loch has shown patchy but often high levels of water contamination of polycyclic hydrocarbon residues.[12] This is bad enough, even without data on, for instance exhaust contamination of biological organisms. Elsewhere in Britain strict speed-limits have been imposed, for example the ten miles per hour limit on Lake Windermere in the Lake District. I understand the LLT Authority has plans to insist on high efficiency engines, so far unheard of in pleasure craft or, better still, efficient engines also running on LPG.

The simplest, most effective move would be to ban motorised pleasure craft; maybe Park designation could do for the loch what apparently could not be done under its combined status as Regional Park, Environmentally Sensitive Area (ESA) and National Scenic Area (NSA) with thirty-odd SSSis, SACs and SPAs on its shores and islands. But as Jack Bisset, Chairman of

8. *Balloch marina on Loch Lomond, Dunbartonshire.*
© Colin McPherson

the Interim Committee (Park Authority-in-waiting) explained to me, the law forbids any interference with free navigation on the loch or on any inland water with a navigable waterway link to the sea. It is a centuries-old piece of case law and any 'responsible organisation' could appeal to get the precedent changed in court. But the Interim Committee feels that the public opposition would be too great to risk, especially as it might lead to a tiresome, expensive public inquiry. How come Loch Lomond is reckoned to have a navigable link to the sea? The answer was that the River Leven may be small and have a barrage across it, but in 1993 someone testified to the effect that a boat can get through. This was accepted, even though the context was as serious as a Fatal Accident Inquiry involving a speedboat.

It is not hard to think up ways in which zoning could be used to manage this kind of problem, for example limiting the space available and/or the vehicle speed. Moreover, the ever-present risk of pollution from normal agriculture and forestry, and from un- or partially treated sewage, can be relatively easily avoided by regulated practices. Recent data from SEPA showed significant phosphorus pollution in Loch Lomond, that is the water is being fertilised (eutrophication) by organic waste (a polite phrase) and artificial fertiliser leakage.[13] The result will be algal blooms, gradual de-oxygenation of the water and hence creeping death to oxygen dependent life-forms in the loch. Other threats to the area are more difficult to deal with: trading in land, planning permission issues and rows about 'appropriate development'. Scottish Environment LINK refers to the 'honeypot' effect of special designations, in other words being called a National Park might alone raise the area's public profile and attract more developments; there is also the 'edge' effect, which could put undesignated land on the Park perimeter under greater pressures.

The LLT Park and its future have been thoughtfully examined by someone who loves the area and worries about it: *Loch Lomondside* is written by John Mitchell, the retired Senior Warden of the Loch Lomond and Ben Lui NNRs.[14] His main cause for concern is wear and tear by tourism. I was surprised that he did not mention the escalation round the corner. On one side of the River Leven, Balloch Castle Country Park is being gently refurbished, which sounds a very positive idea, but on the other burgeons 'a new concept in tourism', a top-rated project called Lomond Shores.

The Lomond Shores project is picture postcard tourism at its most hyperactive. It is enthusiastically backed by the Dunbarton Business Education Partnership. This £60 million complex on a 100-acre site has a visitor forecast of 1.5 million visitors a year.[15] There is a lack of reliable data, even at this relatively late stage. It is not clear whether the forecast means new visitors, or just diverting some of the present flow, allegedly an incredible 5 million who 'pass through the area between May and September'.

The buildings by the Glasgow practice Page & Park look good, all wooden decking and a mock-castle design for the main visitor attraction, four floors of plate-glass sheltered viewing-areas. It has a cinema too ('the newest Extreme Screen™') that will show a 35-minute film about a 'ghostly love story and legend on the shores of the beautiful loch as described in the song'. Good-looking or not, the whole complex – the 100-bed hotel, a coach and car park, the 100,000 square feet of 'sophisticated retailing gallery', the 'luxury residential housing' – is in the wrong place and created for the wrong reason, which is profit for its commercial backers. The main investors are Scottish Enterprise, the European Regional Development Fund and a company devoted to developing 'urban entertainment centres' called The Sheridan Group. It is intriguing that these organisations have got together to invest millions in an enterprise that will keep the visitors away from any disturbing experience of real nature. As the Lomond Shores Brand Values assure us, it will 'mirror the best of the world's leading visitor attractions by creating a new concept in effortless enjoyment and entertainment'.

Picture Postcard: The Cairngorms

'Effortless enjoyment and entertainment' are not the first things one associates with the Cairngorms, but the tourist industry is doing its best to change the rugged image of this iconic area. The main part of the Cairngorms is a wild and dangerous range of high hills that surround the central massif of Cairn Gorm. It is a World Heritage Site but located in the Highlands, which means that it is in an area of economic deprivation with an Objective 1 rating from the European Structural Fund (until 2001). In the Cairngorms, the Scottish tradition of land ownership has meant that in 1990, some 80 per cent of the acreage was in the hands of private owners – the estates of Rothiemurchus, Mar Lodge and Glen Feshie – with most of the rest owned by either SNH (then NCC, about 12 per cent) or RSPB (about 8 per cent) and the Highland Council in charge of the rest.[16] The private holdings are large sporting estates but also provide a mix of countryside tourism, timber production and hill farming. The town and village communities were living off tourism reasonably well, if sometimes gracelessly – with some justice, Aviemore tends to get the worst write-ups. Even the Aviemore Partnership that since 1998 has supervised the refurbishment of the place admits guardedly: 'A demolition squad has moved into the Aviemore Centre to flatten some of the buildings which were built in the 1960s but have failed to stand the test of time.'[17]

The plans for Aviemore are extensive, almost on the scale of Lomond Shores: 'A major redevelopment of the centre, which will complement the redesign of the village as well as the proposed Cairn Gorm funicular railway project, and a new hotel and golf course planned for Dalfaber. . . .' The proposal includes 'marketing opportunities of selling an integrated package of all-year-round attractions . . . Macdonald Hotels [will] also refurbish [three hotels], create three units housing a theatre, cinema and conference centre seating 600 [and also construct] a new arena . . . featuring a multi-purpose sports hall (2,400 m^2), exhibition space (7,000 m^2), a 25-metre swimming pool with flumes, spa etc., a seven-lane curling & skating rink and a gymnasium, dance and fitness studio'. Now this is quite impressive – add to it the casually mentioned Cairn Gorm funicular railway project and the tourism developments in this area shoot up into the seriously large class.

To commit a small Highland town to a glossy resort economy might be defensible if for no other reason than that Aviemore has depended on visitors for over a century, increasingly so as car and coach travel has grown.[18] But the development details should not obscure the most important part of the quote, that is 'as well as the proposed Cairn Gorm funicular railway'. It refers to the project, now under way, to get ordinary tourists, as opposed to skiers, up the slopes and on to the main summit plateau. There is no environmental justification – and it seems, no convincing economic or social reason either – for mauling one of Scotland's most remarkable and celebrated landscapes. The row about the funicular and the additional constructions, especially the eccentric summit café and enclosure, has finally been picked up by the national media after years of negotiation and sometimes open conflict. The main reason that the decision to construct went ahead seems to have been that pressure was exerted by Highlands and Islands Enterprise (HIE; formerly HI Development Board), backed by the Highland Council, on behalf notably of the Cairngorm Chairlift Company and the building contractors, Morrison plc. Building the whole railway/café complex in this difficult terrain is expensive and interestingly, most of the funding (about £15 million in total by the beginning of 2001) has been in the form of grants from the European Union.

The skiing access idea was submerged in a general drive to widen the appeal of these hills. The Scottish Ramblers' Association quotes a Chairlift Company representative as saying: 'It is for the zimmer brigade,' when asked in 1994 why the company wanted a funicular railway.[19] The income from skiing is uncertain and climatic warming makes snow-prospects even more so. One economic solution seemed to be to attract large numbers of coach-borne, less active tourists to a mountainous area that has been the preserve of the reasonably fit.

The development plan caused the previous Tourist Management Programme (TMP) to go into dormancy. The committee framework was provided by the Cairngorms Partnership, a geographically and socially complicated body. Geographically, because the borders of the Cairngorms were set to extend well into Moray, Angus and Perthshire and socially, because of its inclusiveness. The huge membership has been discussed in Chapter 3 and is listed in the Notes.[20] Many feel that the structure of the partnership was crucial in tipping the balance against the environmentalists and their potential allies. SNH was in favour of zoning, but no one seemed able to convince the pro-development lobby that the 'zone of maximum nature conservation' should – must – include both the summit plateau and the sensitive slopes of the mountain itself. For periods other than the skiing season, a compromise TMP directs that in order to 'protect the fragile wildlife on the summit plateau', tourists will not be able to leave the café at the railway terminal.

This plan has consequences that might seem funny, were they not so sad. For instance, it excludes hill walkers from using the railway, because their intention is precisely to stray outside the café enclosure, and so sets up tricky identification and policing problems – who looks likely to make a break for it, and if they do, how get them back? In a thoughtful review of the impact of the scheme and the curious way HIE and Morrison plc seem joined at the hip, Rob Edwards discussed in the *Sunday Herald* the widely-held view that the railway/café project has no strong commercial potential.[21] On the whole, the sporting, environmental and conservation NGOs were in fact curiously muted, in spite of some strong reactions by individuals.

'Curiously muted' because nature conservation in the Cairngorms has been a matter of deep concern not just for the last decade, but since the expansion of tourist provision in the 1960s. The Swedish representative of the international conservation community with a longstanding personal commitment to the Cairngorms, Kai Curry-Lindahl, wrote in 1990: 'The Cairngorms is the largest and most spectacular nature reserve in Great Britain . . . also the most genuine natural area in the British Isles . . . Internationally, it is of significance.'[22] After describing the kind of landscape he was talking about, Curry-Lindahl went on to say: 'The fact that the Reserve was established [in 1954 and enlarged in 1966] for the specific purpose of conserving its native woodlands, mountain habitats etc, makes the devastation of the area over the last twenty years almost incredible.'

9. *Scots pine trees on the Mar Estate on Deeside, remnants of the ancient Forest of Caledon.* © Colin McPherson

A National Park Consultation Exercise

A decade later, the latest (and last?) Scottish Minister for the Environment returned to old non-policies at a meeting called 'Commercial Futures in Cairngorms National Park'. Presumably it was what his audience of business delegates wanted to hear: 'National Parks are not about preserving areas in aspic, creating glorified museums where change is barred. They are about sustainable rural development . . .'[23] It is of course in no way possible to call the developments in those hills sustainable. They consist of at best incongruous and often ugly constructions and buildings, and the hillsides are scarred not only by the Chairlift Company's efforts but also by a 'desperately' worn network of footpaths.[24]

The area has got its own SNH consultation document, *A Proposal for a Cairngorms National Park*.[25] It is less detailed than the Loch Lomond one and does not include any locality assessments. How SNH will be able to complete the proposals without a word of criticism about the landscape destruction is hard to see.

Interview with Mats Olsson and Gunnar Zettersten

(Director & Chief Executive and Member of the Board with special responsibility for Nordic collaboration respectively, at Naturvårdsverket Swedish Environmental Protection Agency)

During a fact-finding visit to Sweden in November 1999, I spent a valuable afternoon at the central Stockholm offices of the Swedish Environmental Protection Agency – Naturvårdsverket (NVV) or the 'Board for Care of Nature'.

Mats Olsson started by sketching the recent history of NVV from its emergence in the forefront of policy-making in early 1980s. This was triggered by public engagement in issues such as PCB (poly-chlorinated biphenyls) pollution, acid rain and pesticide contamination of food. The Department of Agriculture used to be its parent ministry (NVV was created in 1967), but it was transferred to the new Department of the Environment (Miljödepartementet) in 1987. With a widening bureaucratic base came higher profile in Parliament and increasing public interest.

On the issues surrounding National Parks and land distribution, Mats Olsson touched on the traditional patterns of individual ownership, usually as smallholdings or medium-sized farms. Large private estates were relatively rare. The state has been a big purchaser of land, and its holdings are managed by a dedicated agency (cf. the Crown Estates). Most of the companies with large land holdings for mining, forestry and engineering used to be nationalised. Huge stretches of mainly state-owned, very sparsely populated land with the bulk of private ownership in the form of small-holdings permitted setting aside National Park land in the north of the country without much conflict. Other designations were used, for example when the local authority owned the land but accepted that NVV decided targets and regulations. Park creation in the more densely populated south has been a more complex business than in the north, negotiations take a long time and may lead to compromises such as 'island communities' without National Park status, but surrounded by the National Park territory.

Mats Olsson admitted that visitors could cause problems and under certain conditions, access is forbidden (tillträdesförbud). These conditions would include potential for causing damage to wildlife, nuisance to local people or erosion of the land. Whatever the cause, there is no compensation for local businesses but there are arrangements for payments to landowners for restrictions on movement or damage. Objections based on the famous 'allemansrätten' (free access for all) can be overridden by conservation concerns. Recent cases have included forbidding landing on islands in the archipelago to protect nesting birds, closing a ski-run to let the hillside recover and banning jet-skis everywhere except in licensed areas. The coastline protection laws (strandskydd) among other things forbid most building within 100 metres of the beach. In disputed cases, appeals against NVV's rulings (some 10–15 per year) go directly to the ministry. Some local authorities are particularly keen to challenge NVV rulings, usually because they have high-value land for construction or tourism developments. However, NVV gets copies of all planning applications in good time and on past experience, rarely loses appeal cases.

On the subject of road building, Mats Olsson admitted that it can cause controversy, but confrontations of the kind seen in the UK are uncommon. He assured me, with apparent sincerity, that the pro-construction lobbies are always given a fair hearing and that he had always found the state-run transport authority not just very flexible but also deeply respectful of nature. As for air transport, all agree it poses serious environmental problems but so far, the relatively small Swedish airports have not been regarded as first order hazards and re-routing planes has dealt with the noise nuisance. Arlanda, the well-out-of-town main Stockholm airport, is an exception. A fast rail-link between Arlanda and the joint over- and underground railway station in central Stockholm was the latest investment in traffic management. Olsson rather mischievously pointed out that the airport authority had supported the rail-link, calculating that Kyoto-style trading in CO_2 emissions would allow it to run more planes in proportion to reductions in car traffic. Mass air transport is a big issue and NVV is driving the debate round the question 'Should Sweden support air travel?' However, a similar trade-off had been accepted for the new road-link between Sweden and Denmark, the bridge across Öresund Straits: car charges were set with the up-front intention that they should pay for effective and cheap public transport. Every aspect of the bridge had been subjected to wide-ranging Environmental Impact Assessments (EIAs) from both sides of the straits.

A major restructuring of environmental law was completed in January 1999, which saw the publication of an Environment Act (Miljöbalken). Olsson referred with satisfaction to the added back-bone given to conservation by the European regulations, notably in *Natura 2000*,[26] and the increasingly central role of NVV following recent changes in environmental legislation. For instance, although the Swedish Forestry Commission (Skogsvårdsverket) is in charge of setting felling targets, distributing woodland grants and so on, NVV has the statutory right to approve or disapprove of particular schemes. Another key principle is that in court cases, the responsibility lies with the appropriate agency, but the government and in the last resort, Parliament, has final jurisdiction.

Mats Olsson's colleague on the NVV Board, Gunnar Zettersten, began by speaking about Nordic co-operation in environmental matters. The framework for joint working includes the Nordic Council (Nordiska Rådet) for parliamentary politicians and the Nordic Council of Ministers (Nordiska Ministerrådet) for ministries of state and governmental bodies such as NVV. Interestingly, Scotland raised the issue of associate status but got nowhere, presumably because the UK Foreign Office thought its present membership of the union with the rest of Britain would preclude a Nordic affiliation.

There is also Nordic collaboration at the technical level, including on EIAs. Private enterprise is expected to accept regulations based on EIAs, but particularly large, inter-Scandinavian projects can cause protracted arguments. One example was the data 'deemed insufficient by required impact assessment standards' for the natural gas pipeline running eastwards from the Norwegian coast to supply Norway, Sweden and Finland. Another case in point concerned the effects, on land and in the sea, of the electromagnetic fields generated by an electrical cable running to Poland from the south-eastern Swedish coast. The corporations responsible, the Swedish Hydro-Electric Board (Vattenfall) and Southern Power (Sydkraft, now mainly German-owned, though the Norwegian StatOil has got an interest), were told that the field-effects had not been assessed stringently enough and the whole exercise had to start all over again.

Interview with Roger Crofts
(Chief Executive of Scottish National Heritage)

Six months later, I was listening to the Chief Executive of Scottish Natural Heritage (SNH) speaking about environmental management and nature conservation in Scotland. Roger Crofts has been in charge of SNH since its establishment in 1992.

SNH is a remarkable institution: its remit to advise on every aspect of Scotland's natural heritage sounds relatively mild, but in fact takes it into practically every political hot potato in Scotland, including land reform, rural economic development and biotechnology.[27] Roger Crofts gave careful answers to questions about his own and his institution's role in the governance of Scotland. He referred to a variety of expert colleagues when I asked about the use of the law in general and planning law in particular in environmental contexts. Generally, SNH cannot stop things happening, because it has no statutory powers to do so, but can give reasons for or against a particular course of action. It is an important and influential agency, but I did not get the sense of it being in control in the same way as its Swedish counterpart.

Moving on to the National Park legislation, which at that point was just over a month away from being presented to the Scottish Parliament, Crofts did admit a certain weariness with the political machinery, which has a momentum of its own. He argued that the delays which had affected the park proposals until the election of a new devolved executive had been rooted in a profound lack of political will, rather than the efforts of any particular lobby. Even after the new Scottish Parliament had taken over, there was little interest in Scottish Parks until Donald Dewar, the First Minister, came out in favour. National pride and a wish to enshrine Scotland's emblematic landscapes played their roles in this decision, perhaps more so than an impulse towards better nature conservation.

I wondered if there was any contradiction between the spirit of park legislation on one hand, and freedom-to-roam legislation on the other. Crofts felt strongly that responsible access to the countryside was fundamentally non-controversial and long overdue. He did express some reservations about the rank-order of goals for National Parks. It is the policy of the SNH to achieve 'a level playing field' for the competing social and economic activities under the umbrella of sustainability, but only provided that the long-term preservation of natural systems and resources is ensured.

Though admitting that consensus-based guidelines were not always sufficient to deal with all planning issues, Crofts said that they include environmental considerations to a much greater extent than they ever did. Attitudes are changing and although land use is a complex area, in several recent cases SNH has succeeded in getting natural systems recognised in planning guidance.

There is still much concern in SNH about limiting 'Potentially Damaging Operations' (PDOs), but also an open-minded approach to negotiated management practices, for instance in farming. An example is the successful SNH initiative called Targeted Inputs for a Better Rural Environment (TIBRE), which aims to show how technology can be used in farming to benefit the environment and how agri-environment schemes can provide all-round benefit.

On the matter of Sites of Special Scientific Interest (SSSIs), often seen as thorns in the flesh by farmers and other land users, Crofts felt that on whole the system worked well, in spite of losses of valuable sites. These were bound to happen, since in the end the legal fact of property ownership overrules the public interest. Although the landowner is obliged to give notice of intent to interfere with an SSSI, it is not a matter of asking permission, only of announcing his plans with reasonable time set aside for negotiations.

Finally Roger Crofts touched on the funding and enabling functions of SNH. It is becoming increasingly concerned about being seen to be in control of large land holdings and no longer co-funds purchases. Instead SNH gets involved in the management of estates and sites, particularly in collaboration with bodies such as RSPB (for example Abernethy Forest) and the John Muir Trust (for example Ben Nevis).

Interview with Christopher Smout

(Professor of Scottish History and Director of the Institute for Environmental History at the University of St Andrews and Historiographer Royal in Scotland)

Christopher Smout has not only written extensively on environmental history, an important subject he has helped to develop, but has also worked for nature conservation in various roles (he received a CBE for his work in 1994), which included serving as Deputy Chairman of the SNH Board during the early 1990s. His knowledge of the Scottish landscape is unrivalled on many counts, for instance as a historian, naturalist and well-connected environmentalist.

The interview began with a discussion of the many functions of SNH. I reminded Christopher Smout about a passage in one of his recent books, an anecdote describing a significant wobble in the relationship between the newly created SNH and the Conservative administration.[28] When SNH followed its remit to provide advice on how to do things sustainably by pointing out that a second road-bridge across the Firth of Forth would be ill-advised, it was told off for its troubles and threatened with a budget cut. This is a perfect example of the recurring political tendency to double standards. He quoted the Secretary of State's views on the role of SNH: '[A] powerful, independent, authoritative and expert body with advisory and executive functions . . . putting sustainable development concepts into practice'. The use of the word 'expert' led Smout to comment that it must have meant people with expertise in subjects like ecology and environmental statistics. The Secretary of State later called for SNH Board members with 'scientific expertise of the highest calibre'. Smout pointed out that only one of the twelve Board members was a scientist during the first seven years of the SNH's existence, even though they were of course appointed by the Secretary of State. There is no question that nature conservation suffered from the lack of scientific expertise on the Board. He quoted not without relish a dictum used at the foundation of SNH by the representative of the Scottish Landowners' Federation[29] 'Better have the scientists stay in their universities', adding that this seems to have been a sentiment shared by successive governments.

Maybe the combination of 'independent' and 'statutory agency' is inherently contradictory. So were the goals set for SNH by the Conservative administration: it was to deal with what was seen as a 'loss of support for conservation' in a climate favouring economic development and 'limit any such conflict as much as possible'. Here SNH has been more successful than many might have expected. Smout argues that SNH has built up key skills in managing partnerships. 'Access to the land' is one example. The debate was channelled by SNH into a genuinely collaborative forum and led to practical agreements. Such group consultations can, if well handled, be productive and helpful. There have been other instances of co-operation, from national to strictly local, that more than vindicates an integrated approach to the proliferating culture of interest groups. Wide-ranging consultation is an important element in interest group politics, but can turn out hard to handle for people without the right skills. Zoning is another form of 'constructive conflict resolution', but the emphasis is still on jobs and economic development.

Christopher Smout did not see the ambition to 'have it all' – both protection of wild land and dynamic economic development – as a merely present-day goal but traced it back to post-war planners such as Robert Grieve. As a nature conservationist, Smout was not fond of the National Park structures that have developed in Britain. Both the English and Scottish versions seem unable to deliver the benefits of nature protection, while at the same time irritating local people with complicated by-laws. He regretted the lack of clear, enforceable planning guidelines and the fact that the principle of precedence for nature conservation in cases of conflict (the Sandford principle) will almost certainly be lost in the final Scottish legislation. Still, at present National Park designations may well be the best way of managing large areas of Scotland's landscape. The emphasis should lie on the fact that these areas are of national importance. It may be good democratic practice to prioritise the concerns and ambitions of local people, but a national resource, funded by the nation, should be managed for the benefit of all.

I worried away at the sheer number of environmental NGOs and voluntary organisations. Smout said that the Scottish Environment LINK is an effective umbrella body, but also that generally all its member organisations have crucial contributions to make. In conflicts between local people and allegedly narrow-minded, dictatorial interest groups, Smout tended on the whole to support the latter. The onus was on the wildlife charities to explain themselves well and work with the grain of local community feeling, rather than trying to widen their goals so far as to become rural development charities. Protection for wildlife should be seen to help the communities, as is the case with the Scottish Wildlife Trust's presence on Eigg. Another example was the RSPB's attempts to introduce corncrake-friendly land management on Coll and elsewhere, although there have also been instances of apparent, though unintended RSPB arrogance. The 'oats for black grouse'-story was nice example of collaboration. Listening to Speyside people unearthed the fact that in the past, when black grouse were more common, they used to turn up in fields where oats were grown. Reasonable payments for growing oats and harvesting them in the traditional stooks produced benign results, as both the farmers and the black grouse (still very rare) enjoyed the return of the long-lost crop.

Conflicts often produce bad-tempered calls for 'the scientific evidence'. This is both reasonable – of course decisions should be evidence-based – and unreasonable – what you do with the evidence is more subjective than objective. The present systems for collecting and analysing data are insufficient and the rows about rights and wrongs give credence to those who want to keep scientists from influencing decision-making. Christopher Smout said that the nature and wildlife charities do invaluable work on data collecting and on monitoring species increases and declines. SNH does not have the resources to cover all the monitoring it should do and often relies on the work of these enthusiastic volunteers, which can be outstanding as, for instance in the case of the British Trust for Ornithology. Its volunteers have collected and the BTO scientists collated data to the standard that has been called 'citizen-science' in the USA.

In a fascinating aside, he spoke of his own views about non-native and native species. Although it is fashionable to prefer native species, it makes no sense to label non-native species 'alien' and try to exclude them, if the only grounds are that they might interbreed with native species. This is especially true if protection is refused in Britain when a species is under threat elsewhere. He admitted that some non-natives could drive out native competitors or damage habitats, and that both non-native and native species can be pests or 'vermin'.

The role of hunting, shooting and fishing in countryside management raises another difficult set of issues involving conflict-prone interest groups. Given that Smout is a committed naturalist with bird-watching high on his personal agenda, I felt this might be a controversial subject. However, he seemed quite cheerful about the possibility of naturalists and hunters getting on in peaceful co-existence. This is another area where partnership-based negotiation has been fruitful. The British Association for Shooting and Conservation (BASC) runs excellent nature reserves, including Mugdrum Island in the River Tay. It is 'the best one in Fife' apart from the Isle of May. BASC, moreover, once helped to defuse what could have ended up as a nasty confrontation: the arrivals of Italian bird-shooting parties with a penchant for huge game bags (often including protected species) and over-powerful guns (pump-action rifles) caused real hostility among local people. BASC was able to talk 'shooter to shooter' and, with the help of SNH, prepared useful instructions about local rules and regulations in Italian.

Game management for sport provides significant income for

many members of the Scottish Landowners' Federation (SLF), an efficient and powerful pressure group that is eyed with suspicion by many conservationists. Smout said that quite a proportion of SLF members are 'green' by any standards and that the organisation's increasing emphasis on 'stewardship of the wilderness' is not only just clever window-dressing. However, in spite of the organisation having an environmental subcommittee, the internal policing of the not-so-benign practices of any members who are unsympathetic to conservation is not guaranteed to be effective.

When free public access is enshrined in law and regulated by the SNH-produced Countryside Code, policing is likely to become a bigger countryside issue. It is going to be problematic both to make people aware of the rules and to enforce them. The rule of law in the countryside can present much thornier problems still. The English Countryside Bill aims to improve the protection of sensitive areas such as SSSIs and Nature Reserves (NRs). Smout pointed out that the great majority of designated conservation areas are in no sense 'owned' by the public. Because their protection depends on the landowner, they can be very vulnerable. Nature's status under European law is better than under Scottish law. Ken Collins, the Chairman of SEPA and with years as Chairman of the European Parliament's Environmental Committee apparently once said that he had a rule-of-thumb international rank order of environmental protection: Scotland, he reckons, is way behind Scandinavia and the Netherlands, and in many respects England, but better than Ireland and much better than Spain.

A country's relationship with nature depends on factors such as the degree, speed and nature of its industrialisation; the ratio of land to population and laws of land ownership all change. Smout felt that the main reason for the speed with which land reform is moving up the agenda of the Scottish Parliament may lie in its symbolic importance to people in Scotland, rather than because of any changes in land ownership or management it will bring about. The Free-to-Roam campaign too had symbolic significance as the end of much-resented exclusion. He added that the freedom of the hills should not be extended to dogs off their leash. It was essential that the right to roam should be limited with a leash-law for dogs in the countryside – though it was hard to imagine that restricting the freedom of British dogs would go down well in a Scottish Parliament.

Recreation and country sports can all lead either to healthy, happy communion with nature or to danger and destruction. The gentle art of canoeing at the wrong time, and/or practised by too many people in the same place, can cause damage to sensitive lochs and their populations of animals and birds. Rather than sport everywhere, he spoke warmly of Rutland Water and Grafton Water, two English examples of combining nature conservation and recreation, including water sports. There is a similar success in Fife at Lochawe Meadows, a wetland area created on old mine workings and now imaginatively run by the local authority. All depend on zoning with some areas set aside for sport and some for nature conservation.

Scottish Land Reform: Land or Landscape?

At the time of writing, land reform-related stories are spilling out of a bulging file. Material on nature conservation policies such as National Parks and SSSIs is stored elsewhere but it is still hard to see a clear pattern in the remaining case histories. The most obvious factor in common between the arguments surrounding the Land Reform Bill on one hand, and the steady flow of cases and conflicts with land at their centre on the other, is the role of landowners with large or very large holdings. The rights of local people predominate in a way that has a long history in Scottish public debate. 'Sustainable land use' is not a key issue and seems to mean anything and everything.

It is strange that the legally sophisticated Scottish nation should have ended up with a system of land ownership so ancient and at least in part so unfair that the local variants of it were scrapped elsewhere in Europe long ago. The possibility that it has conferred unique advantages in the Scottish setting, for instance with regard to 'stewardship of the land', must not, however, be forgotten in the chorus of condemnation. There has been a succession of attackers, deeply engaged, articulate and well-informed – mostly through their own patient research – whose work has led directly to the devolved Executive's perhaps greatest piece of legislation to date, the Land Reform Bill (still to be enacted in full at the time of writing).

The real start of their campaign came in 1977 with the publication of a small book called *Who Owns Scotland?*, written and compiled by the then 90-year-old John McEwen.[30] A social radical, he had worked devotedly to find an answer to the question posed by the title of his book, and the result, while not entirely accurate, was crucial. The next steps were taken by much younger men, notably Robin Callander, James Hunter and Andy Wightman, who wanted political victories for Scotland's land-starved population. Robin Callander handed the 'who-owns-what'-chase on to Andy Wightman, who wrote *Who Owns Scotland* (no question-mark needed any more), twenty years after McEwen. It is a powerful combination of research data and polemics.[31] Callander himself turned to the legal issues of landownership in Scotland – the feudal structure, the role of Crown supremacy, public ownership and public access, tenancies, leases, communal ownership – in a book on *how* Scotland is owned, that manages to be both complex and lucid.[32] As the Reform White Paper was being turned into real legal change, Wightman spoke at the annual McEwen lecture in 1999 and published a book, both partly triumphant and partly severely critical of some of the new provisions.[33, 34]

Social justice has been the reformers' main goal. That means rights for tenant farmers to become free-holders and use the land in their own best interests, and limitations to power of the few who own so much of Scotland. About 0.025 per cent of the population own 65 per cent of all rural land, an extraordinary figure.[35] Because the debate has focused on a fairer division of the land and greater scope for rural development, the effects of the reforms on the landscape have taken second place. Still, Andy Wightman refers to an 'environmental discourse' and at one point quotes Aldo Leopold, the famous US environmentalist, on the danger of treating land as a commodity instead of '[seeing] the land as a community to which we belong, [so that] we may begin to use it with love and respect'.[36]

'The environment' tends to be brought in as an afterthought by both sides, though. For of course, there has been another side, led by the Scottish Landowners Federation (SLF), with lower public profile but well developed lobbying skills. SLF has lost many battles, but whether it will have lost the war remains to be seen. It is not easy to find economic or social justifications for very large private landholdings, and perhaps particularly for the Scottish speciality, the high-priced but untaxed wild land known as sporting estates.

The Land Reform Bill of course has many sections dealing with access and land use. So many, in fact, that focused campaigners like Andy Wightman have become annoyed: 'A great many topics, some of which have nothing to do with land reform at all . . . for example recommendations on National Parks, codes of good practice on land use, and community planning'.[37] The Scottish Executive's reasonable view is that these are crucial aspects of the legislation.

Some of the best arguments from the SLF lobby put the environment first. Owners like the Grants of Rothiemurchus have turned their estate into 'an icon of enlightened conservation management'.[38] An extensive report on the economics of the sporting estate that is still valid also concluded that sporting proprietors almost universally regarded themselves as conservationists; respondents felt that sporting use has a better (nature) conservation record than forestry, agriculture or letting the ground go wild.[39] There are other arguments for the old forms of estate management. For an owner with deep roots in the land of his/her origins, the privilege of ownership is likely to be matched by a willingness to invest in traditional values. These might be seen as outmoded, but imply life-long understanding of the countryside and the people who depend on it. Michael Wigan set down and elaborated these ideas with real conviction in his book about landed Scottish gentry with the helpful title *The Scottish Highland Estate. Preserving an Environment.*[40] The only serious problem with his argument in favour of the status quo is that thanks to the poor legal controls of land purchases, land use and conservation, there are too many bad eggs in that particular basket.

It may be that the overtone of 'a better class of people' – 'gentry' – present in this line of thought causes the resentful reactions to it. Still, in an old country with great pride in its history, it is not unreasonable to present traditional values and practices as models for future developments and continuity as part of their strength. An outsider might assume that the other traditional British concern, social class, is somehow infiltrating the urban versus rural debate. The *Private Eye* 'Letters' pages have been reflecting, as they always do, current thought-patterns by large numbers of abusive messages passed between supporters and detractors of the Countryside Alliance:[41]

> Your advertisement for the Countryside Alliance shocked me. How can we be against multi-millionaires buying the Conservative Party when you are prepared to be bought also.

> All these 'right-on' creeps (aka Disgusted of Islington, aka Tutting Puritan against Yokels, aka Fluffy Bunny Lover etc) supporting fashionable 'liberal' causes, whinging that the Countryside Alliance should dare to advertise its views . . .

There are rational arguments against the 'traditionalist' case, especially if one focuses on the environmental issues. This eliminates the often unverifiable claims and counter-claims about the social and economic functions of the large estate as employer and investor in the countryside. The unregulated market in land has meant that a large proportion of Scotland is not owned by British citizens living on the land. With tight enough regulations, management by factors or foreign company ownership need not be bad for the environment, but it does mean that the reliance on 'local roots and traditional values' must be replaced by detailed legislation of just the kind landowners of any size tend to mind very much.

Ian Mitchell's book *Isles of the West* is a polemical tract about nature conservation interfering with 'local folk', though lightly disguised as an idyllic travelogue.[42] Mitchell is for Everyman (mostly) and against officialdom, ranging from tolerable (National Trust for Scotland) to tolerable but possibly inefficient (John Muir Trust) to the unbearable, possibly dishonest

(Scottish Natural Heritage, Royal Society for the Protection of Birds). It is a strange book, worth reading for the intelligence and passionate self-belief of its author, but a failure as an argument because of its inconsistency. His distaste for people 'from Edinburgh' who turn up on the doorstep and tell the locals not to shoot geese (Mitchell lives on goose-infested Islay), to be nice to corncrakes and otherwise do what they're told seems to be based on having met a handful of tiresome, possibly ill-informed conservationists. The conflict lies between his laissez-faire attitude towards local people and his love for the nature that the locals on more than one occasion have been prepared to sell to the highest bidder.

The nature conservationists are land-hungry and can seem imperious, but are unlikely to trade in environmental assets. Their advice to local communities, once freed of accusations of alienation and overbearing manners, tend to amount to things like running farms a little less efficiently for the sake of birds (RSPB), clearing up rubbish, planting trees and looking after SSSIs by the book (SNH) and excluding certain forms of tourist development from their estates (John Muir Trust). Mitchell found one example of the ecological approach in the set-up on the Isle of Rhum, because it has got no 'local people' in his sense, only incomers who are conservationists or tourists with outdoor interests. His criticism here was typically people-oriented: the tourist-potential is great but the SNH-managed landscape is crawling with boring scientists who should find something better to do.[43]

But the book also contains accounts of the locals being demanding and sometimes destructive of the landscape. He is sharply critical of the orthodox policy to 'sustain rural communities'. The subsidy-hungry but conservationist community that bought Eigg is not too different from the cash-starved bar owner on Harris who hopes against hope to sell his granite hill for construction industry aggregate.[44]

Landownership confers rights to buy, sell and 'develop' on the owners; it would seem that one form of ownership is little different from another when its comes to putting earning more money ahead of love of the land. My collection of land-based stories include many featuring arrogant and grasping landowners.[45] The biggest recent case of a 'bad laird' interfering with the heritage of the rest of the Scottish community is the marketing of the Black Cuillin on Skye by the chief of the clan McLeod. John McLeod seems to have won the legal argument and some sympathy for his need to shore up his castle, crumbling for more than a generation and a tourist honey-pot. This seems a case where the nation should decide to make a conditional purchase of the mountains, if for no other reason that wilderness on such a scale is too precious to be traded.

No landowner can be trusted always to put the welfare of the landscape first. The new land ownership legislation has become mixed up with land use and management, and while the causes seem more than reasonable – access and conservation – they could be counterproductive. In particular this is true for the prioritisation of community ownership, which is somehow understood to be so generally beneficial as to deserve advantages when it comes to bidding for and purchasing land.

For some, it is an article of faith that community ownership is a good thing, but the recent land use stories have posed hard questions. If you transport an island mountain away as crushed construction aggregate and this supplies the island people with an income and possibly local jobs, is that good? If a whole island mountain range is sold in the open market and this supplies the owner of a historic country house with an income and possibly local jobs, is that bad? Should an island community of about twenty families receive public funding at a rate of about £50,000 per household in order to buy their island? What is the virtue or otherwise of communities owning land? What land, anyway? It is not clear, for instance, if community ownership of urban or suburban land will be supported in the same way as the rural variety.

As both Robin Callander and Andy Wightman have pointed out, distinctions must be made between land ownership reform, land distribution and land use. For land use purposes, the public interest must be defined: 'To retain a direct proprietorial interest in the land tenure system by us the people . . . is important in terms of conditionality, and social and environmental duties and responsibilities'.[46] This is a line of thought that has for instance been followed by advocates of an overriding national interest in National Park management, as opposed to giving priority to the wishes of local people. Interestingly, SLF has decided to come down in favour of the view that: 'we are less happy . . . that communities have to be given the ability to buy land to the exclusion of any other type of purchaser; nor can we accept that groups are given a right to acquire crofting land whether or not the owners wish to sell under a process that is little better than expropriation' and 'Properly directed, an alliance of community groups can do much good'.[47]

This rings alarm bells for those who feel the time has come to give the environment top priority and who have learnt to regard talk of 'development directed by the local community' with suspicion. Andy Wightman, for one, has sharply criticised some aspects of both recent cases of community land purchases and the proposed community 'right to-buy' legislation. Most of the recent deals – involving crofting land at Coigach and Assynt, the island of Eigg, and the already diminished and run-down wilderness of Knoydart – have been made possible by sympathetic media coverage, political pressure to reduce the market price and large cash donations, all of which occurred because in each case the community was seen to be deserving. People did not own their own land, could have their homes bought and sold over their heads, and existing regulation of estate management had not prevented environmental and economic decay. These deals were fine in the short term, but justifying the changes in the law to prioritise pre-registered bids by communities is still not easy to defend. Group purchases of this kind are also bolstered by the new Land Fund (£10 million), under the umbrella of the UK New Opportunities Fund and financed mainly by the Heritage Lottery Fund (see Chapter 8). The Land Fund is earmarked for communities – 'individuals' are mentioned, but the language gives a strong impression of favouring communities.

There are problems of fairness and the future control of what happens to the land. As Andy Wightman asks:' [Are the new rights for] those who live and/or work on the land . . . or should they be aimed at wider communities of geography and interest?' The best reported cases of community purchases of estates, Eigg and Knoydart, would have been unable to take advantage of them because of the [proposed] narrow definition of community and the requirement in partnership arrangements that the community body possess the majority of votes.'[48] What I have seen written by the communities themselves tends to be about hard-won triumphs over injustice, rather than detailed management plans.[49] However, there are hopeful examples from places abroad with long traditions of community-managed enterprises, which will be looked at later in the context of the 'economics of the environment'.

Landscape and Environment

Many problems affecting the 'stewardship of Scotland's landscapes' will not be solved by the current proposals for land reform. Who plans and then enforces environmentally sustainable development of the countryside? How will the accumulating wealth of research data on land use and evaluation of land resources be used? Over ten years ago, in a fascinating collection of papers on 'Evaluation of Land Resources in Scotland', one of its editors wrote: 'If the last ten years have seen land evaluators reacting to rapidly changing policies, perhaps it can be hoped that in the next decade policy formulators will increasingly use the data bases that are currently being developed to create models against which to test their own or received ideas.'[50] This would be an important widening of the public debate, apparently stuck with a 'wicked landlord versus hard-done-by tenants' scene, to include data-based projections of environmentally sustainable land use.

The economic options for landowners are going to dominate that debate, but the landscape need not be excluded. The social anthropologist Tim Ingold says in his *The Perception of the Environment*:[51] 'There can be no radical break between social and ecological relations; rather the former is a *subset* of the latter . . . The first step . . . would be to recognise that the relations between human beings and their environments are not confined to a domain of 'nature', separate from and given independently of, the domain in which they lead their lives as persons.'

10. *The Kyle of Tongue, Sutherland.*
© Colin McPherson

Notes

1. *The National Park the Locals Stopped* by Peter Hanneberg and Gunnar Zettersten. *Enviro* 1994, 17: 16–18.
2. *Urban Planning Theory since 1945* by Nigel Taylor, 1998, p. 2.
3. *Nature Contested. Environmental History in Scotland and Northern England since 1600* by T. C. Smout, 2000, pp. 142–72.
4. www.scotlink.org/np Scottish Environment LINK is an umbrella organisation for some thirty-eight NGOs with environmental interests; for a full list of member organisations, see Chapter 3.
5. *Nature Contested. Environmental History in Scotland and Northern England since 1600* by T. C. Smout, 2000, pp. 162–3.
6. e.g. Robin Harper, MSP (Green Party;); Andrew Batchell (Scottish Environment Link); Frank Bracewell (National Park campaigner).
7. *A New Way of Caring for a Special Place – Proposal for a Loch Lomond & Trossachs National Park. Consultation on the Area, Powers and*

Representation of the Proposed National Park. Scottish Natural Heritage, November 2000.

8. Ibid., 'Key Issues for Consultation: Area', item 7, p. 27.
9. Ibid., 'Sub-areas of Primary Consideration', p. 50.
10. Scottish Wildlife Trust: Peatland Campaign. www.swt.org.uk
11. Stirling (largest area), Argyll & Bute, West Dunbartonshire and Perth & Kinross Councils.
12. 'Hydrocarbon emissions from Boat Engines: Evidence of Recreational Boating Impact on Loch Lomond' by Mark Bannan et al. *Scottish Geographical Journal* 2000, 116(3): pp. 245–256.
13. 'Pollution Alert as Loch Water Quality Suffers' by James Freeman. *The Herald*, 30 October 2000.
14. *Loch Lomondside: Gateway to the Western Highlands of Scotland* by John Mitchell, 2001, e.g. pp. 195–214.
15. Lomond Shores Project. www.loch-lomond.uk.com/lomond-shores A Major Development is Taking Place on the Shores of Loch Lomond. www.dunebp.co.uk/lomond
16. 'The Role of the Nature Conservancy Council in Protecting the Cairngorms' by E. M. Matthew, in *Caring for the High Mountains: Conservation of the Cairngorms* by J. W. H. Conroy et al. (eds.), 1990, pp. 30–49.
17. Vision for Aviemore. www.highland.gov.uk/plintra/complan/focusapr98/aviemorevision
18. 'In Search of Wilderness, Nature and Sport: The Visitor to Rothiemurchus 1780–2000' by Robert A. Lambert, in *Rothiemurchus. Nature and People on a Highland Estate 1500–2000* by T. C. Smout and R. A. Lambert (eds.) 1999, pp. 32–59.
19. 'Moving Mountains: The Cairngorm Funicular' by Rob Edwards. *Sunday Herald*, 31 December 2000.
20. The list has been based on www.cairngorms.co.uk: Scottish Executive Rural Affairs Department. *Councils*: Aberdeenshire, Angus, Moray, Perth & Kinross, Highland; *Quangos*: Forestry Enterprise & Commission, Deer Commission for Scotland, Historic Scotland, SEPA, SNH, SH, SS, STB, Scottish Enterprise, HIE; *LECs*: Grampian, Moray, Badenoch & Strathspey, Tayside; *Charities*: National Trust for Scotland, RSPB; *Others*: Aviemore Partnership, Association of Deer Management Groups, Crofters Commission, The Crown Estate, North of Scotland Water Authority, National Farmers' Union, Scottish Landowners' Federation; *Secondary Partners*: Local Tourist Boards – Aberdeen & Grampian, Angus & Dundee, Highlands of Scotland, Perthshire – Angus Glen Business Group, The Armed Forces, APRScotland, Association of Scottish Ski Areas, Botanical Society of Scotland, BASC, British Association of Nature Conservationists, Cairngorms Chairlift Company, The Cairngorm Club, The Cairngorms Campaign, Cairngorms Community Circle, Cairngorms Mountain Rescue Association, CoSLA,

Council for Scottish Archaeology; *Fishery Boards* – Don & District, Dee, Esk, Spey District, Tay – Farming and Wildlife Advisory Group Scotland, FoE Scotland, Flying Scot, The Game Conservancy Trust, Glenshee Chairlift Company, The Heather Trust, Highlands Birchwoods, Institute of Mountain Environments, Institute of Terrestrial Ecology, LANTRA Trust, Macaulay Land Research Institute, Millennium Forest for Scotland, Mountaineering Bothies Association, Mountaineering Council Scotland; *National Farmers Union* – Atholl, Blairgowrie, Dee & Donside, Forfar, Strathspey, Tomintoul & Glenlivet – Native Woodland Advisory Panel for Scotland, North East Mountaineering Trust, Popular Flying Association, The Ramblers Association, Reforesting Scotland, Royal Scottish Geographical Society, Spey Research Trust, Scottish Arts Council, Scottish Conservation Projects Trust, Scottish Countryside Activities Council, Scottish Countryside Rangers Association, Scottish Crofters Union, Scottish Environmental Education Council, Scottish Environment LINK, The Scottish Executive Education Department, Scottish Gamekeepers Association, Scottish Gliding Association; *SLF Regions* – Central/Highland/North East – Scottish Mountaineering Club, Scottish Museums Council, Scottish Ornithologists Club, Scottish Scenic Trust, Scottish Wildlife Trust, Snowsport Scotland, Tayside Raptor Study Groups, Timber Growers Association, Voluntary Action in Badenoch and Strathspey, WWF.

21. 'Moving Mountains: The Cairngorm Funicular' by Rob Edwards. *Sunday Herald*, 31 December 2000.
22. 'The Cairngorms Natural Nature Reserve (NNR): The Foremost British Conservation Area of International Significance' by Kai Curry-Lindahl, in *Caring for the High Mountains: Conservation of the Cairngorms* by J. W. H. Conroy et al. (eds.), 1990, pp. 108–19.
23. 'Conservation News; Cairngorms National Park, in *Scottish Environment News* by Michael and Sue Scott (eds.), February 2001, p. 1.
24. 'Urgent Pleas for Repairs as Cairngorm Paths Erode' by John Ross. *The Scotsman*, 25 August 1999.
25. *A Proposal for a Cairngorms National Park. A Consultation on the Area, Powers and Representation for the Proposed National Park.* Scottish Natural Heritage, 2000.
26. The Conservation of Natural Habitats and of Wild Flora and Fauna (Directive 92/43/EC). The European Union is in the process of developing a European environmental network called Natura 2000, consisting of the sites and species of interest to the community. It will ensure a protected status for the plant and animal species that are threatened. In the Habitat Directive is mentioned that each member state of the EU is responsible for the conservation, according to Annexes I and II of the directive. The establishment of the Natura 2000 network of Special Conservation Areas (SCAs), as protected areas will be called from now

on, will be completed by 2004 and will be based on the national lists of habitats and species of community importance, that each member state has to submit to the EU. virtuals.compulink.gr/stps/natura2000.html

27. Natural heritage includes nature and landscape protection and enhancement, environmental education, informal access to the countryside for recreation and enjoyment, and sustainability (Natural Heritage (Scotland) Act 1991. The Act refers to SNH's role as advisor to the Scottish Executive and in the planning system. It is funded by the Executive (approx. £50 million per annum) and is not allowed to earn substantive income, for instance from grants, land-holdings or other assets.
28. In 1990, the Secretary of State for Scotland, Malcolm Rifkind, announced a new agency, SNH, replacing the Nature Conservancy Council and Countryside Commission for Scotland. See *Nature Contested. Environmental History in Scotland and Northern England since 1600* by T. C. Smout, 2000, pp. 168–9.
29. P. T. Gordon-Duff-Pennington.
30. *Who Owns Scotland?* by John McEwen, 1977.
31. *Who Owns Scotland* by Andy Wightman, 1996. Also, *Who Owns Scotland Now? The Use and Abuse of Private Land* by Auslan Cramb, 1996.
32. *How Scotland is Owned* by Robin Callander, 1998.
33. 'Land Reform: Politics, Power and the Public Interest' by Andy Wightman. The 1999 McEwen lecture. www.caledonia.org.uk/land/lectures
34. *Scotland: Land and Power – the Agenda for Land Reform* by Andy Wightman, 1999.
35. Ibid., p. 29.
36. Ibid., pp. 48–9. 'Aldo Leopold was one of the early leaders of the American wilderness movement [...] instrumental in the 1924 designation of the first Forest Service wilderness on the Gila National Forest in New Mexico. In later years Leopold developed eloquent arguments for the importance of wilderness preservation, development of a land ethic, and an understanding of the importance of the integrity and beauty of nature.' www.wilderness.net/leopold/intro
37. Ibid., p. 68.
38. Foreword by Magnus Magnusson, in *Rothiemurchus. Nature and People on a Highland Estate 1500–2000* by T. C. Smout and R. A. Lambert (eds.), 1999, p. viii.
39. *The Economic Impact of Shooting Sports in Scotland.* A Report by the Fraser of Allander Institute at Strathclyde University for the Scottish Development Agency and the British Association for Shooting and Conservation, 1990. See ref. 40, pp. 152–3, in the report.
40. *The Scottish Highland Estate. Preserving an Environment* by Michael Wigan, 1991.
41. Letters by J. Raynor and 'Russ Inurbis'. *Private Eye* (1023 & 1024), 2001.

42. *Isles of the West. A Hebridean Voyage* by Ian Mitchell, 1999.
43. Ibid., pp. 58–80.
44. Ibid., pp. 209–14 (Eigg) and 148–50 (Harris).
45. 'Glenshee Farmer Defeats Laird's Bid to Oust Her from Land He Wanted for Shooting' by Graham Ogilvy. *Scotland on Sunday*, 20 June 1999; 'Duke Cared More for Shoot than Staff Welfare' by Frank Urquhart. *The Scotsman*, 2 December 1999; 'Top Scots Landowner Fights Order to Quit Site. Bizarre Dispute Delays Path Completion' by Frank Urquhart and Stuart Nicolson. *The Scotsman*, 24 August 1999; 'Laird Threatened Island Lifeline' by Torcuil Crichton. *Sunday Herald*, 4 February 2001.
46. 'Land Reform: Politics, Power and the Public Interest' by Andy Wightman. The 1999 McEwen lecture, Part I: 'The Public Interest'. www.caledonia.org.uk/land/lectures
47. Draft Land Reform Bill, February 2001; SLF Welcome For Community Land Alliance, June 2000 (Press Releases) www.slf.org.uk/login/slf/press.asp
48. 'Land Reform: Politics, Power and the Public Interest' by Andy Wightman. The 1999 McEwen lecture, Part II: 'Handling Land Reform'. www.caledonia.org.uk/land/lectures
49. *We Have Won the Land* by John MacAskill, 1999.
50. 'Past and Future Decades of Land Evaluation in Scotland' by M. F. Thomas, in *Evaluation of Land Resources in Scotland* by J. S. Bibby and M. F. Thomas (eds.). Proceedings of the Royal Geographical Society Symposium, University of Stirling, 1989, p. 3.
51. 'Hunting, Gathering as Ways of Perceiving the Environment', in *The Perception of the Environment: Essays in Livelihood, Dwelling and Skill* by Tim Ingold, 2001, p. 60.

7
Greening the Mean Streets?

Urban Scotland

URBAN SCOTLAND *is* Scotland, no matter how often the rural landscape is projected as its true source of identity. In the year 2000, about 80 per cent of the population lived in towns of 20,000 or more inhabitants. All but a comparative handful of Scottish children grow up in 'built environments' and, unlike the Scandinavians, Scottish families do not normally have holiday homes in the countryside. The alienation from the reality of the lochs and hills could hardly be more complete. Urbanisation is a world-wide phenomenon, resulting both from declining rural economies and the economic dynamism of the cities. Scotland started early: in the century between 1850 and 1950, the distribution of population swung round to its present pattern. The trend is still continuing, but more slowly. The speed of this change and the industrialisation that drove the whole process resulted in 'The Tenement City'. As T. C. Smout says of conditions in the 1950s: 'The true grimness of the Scottish town [was] concealed from the tourist round of monuments and shopping centres.'[1] That is still true, although the 'grimness' is to some extent generated by different forces.

The great cities have their emotional champions and many small towns can draw on the affection that a familiar neighbourhood creates. Chapter 2 on land- and cityscapes in fiction helped to prove – if proof was needed – that love of a place is not always a function of how highly it rates for conventional beauty or comfort. It is in fact hard to think of a single form of urban existence that has not been a source of identification: small town and suburbia, high- and low-rise, gracious living in decorous squares and fast living in noisy flats. In Scotland, Glasgow and Edinburgh have been chronicled with more adoration than might have been good for either city. Less glamorous cities such as Aberdeen and Dundee can also drum up civic pride. Yet Scotland's urban landscapes have been subjected to depressingly bad planning and design over the last century, perhaps especially in the environmental sense.

How to House People in Cities

Cities are a favourite topic for research and discussion: architects, planners and sociologists top a long list of more or less professional urbanists. There are university departments and research units devoted to cities and a steady output of journals, books and reports. Post-war Europe saw huge changes in city structure, driven by a combination of the familiar shift from rural to urbanised industries, population growth, mass-manufactured architecture and a fashion for vigorous town planning – 'zoning' and suburbanisation in the wake of road-building. The same forces created the world-wide urbanisation that has reached alarming heights in places like Tokyo-Yokohama (31.6 million inhabitants) and Mexico City (18 million). Some of these vast built aggregates are monsters of inhuman ugliness, poverty, pollution and waste.[2] More than twenty years of The United Nations Centre for Human Settlements, now revitalised as Habitat II, has barely chipped at the margins of the compound problems of the 'Metacity'.[3]

It would be a mistake to assume that the Metacity phenomenon has nothing to do with the UK in general and Scotland in particular. The extraordinary, unstoppable invasion of housing into the English countryside is creating urban scatter just like the 'Randstad', an earlier and more extreme version in The Netherlands, which is one of the world's most densely populated countries. The growth of Geddes's 'Clydeforth', the merging cityscapes of Scotland's Central Belt, is another example of the same process. On the other hand, there are striking differences between the responses to urbanisation by the Dutch and the Scottish (or British) architectural and planning establishments.

In 1999, the Dutch MVRDV group produced the incisive, innovative and stunningly well ilustrated *Metacity Datatown*.[4] The same year saw the publication of *Area Regeneration*, the first issues of the *Social Inclusion Research Bulletin* from the Central Research Unit (CRU) at the Scottish Office.[5] It is of course true that the CRU is wholly dependent on the state and MVRDV is at least partially independent (members of the Dutch group have been in and out of state organisations). Unfair or not, it remains hard to think of any excuse for the programme allegedly intended to 'Open the Door to a Better Scotland' by pulling together projects from 'Criminal Justice, Education, Housing, Local Government, Rural Development, Social Work and Transport'. It is symptomatic that the compilers have had to rely on the alphabet to provide a measure of order in the list. The lack of any kind of unifying principles means that the entire exercise appears pointless. What is the outcome of 'New Life for Urban Scotland – Final Evaluation', for instance? The team decided to 'consider lessons learned and look forward to the sustainability of the achievements'. It should have been fascinating to learn about the project 'Cities, Competitiveness and Cohesion in a Scottish Case Study' and the extent to which it contributes to our 'understanding of the factors influencing [the] economic performance of cities and smaller urban centres', but the conclusions are unclear.[6]

Clone City is a more interesting Scottish counterweight to *Metacity Datatown*.[7] This thoughtful book by Miles Glendinning, an architectural historian, and David Page, a leading Glasgow-based architect, was also published in 1999, and also focuses on continuing urbanisation and the forces that drive it. There, easy comparisons end. The contrasts are brought home by the illustrations: colour-coded computer abstractions in *Metacity* and careful black and white 'architectural' photographs in *Clone City*.

The Dutch book has a consistent theme: the human environment is a finite resource. The choice is ideological, but the treatment based on hard data – hence the *Datatown* of the title – and the presentation depends on the computer simulations. The data is – data: the amount of CO_2 in the exhalations of an aggregation of so many people, the number of trees needed to absorb that quantity of CO_2, the number of wind-generators (the least invasive, most efficient renewable energy source) needed to supply Datatown with energy, and so on. There is some hard-hearted humour too. No environmentalist would fail to smile (bitterly, maybe) at the Datatown area for recreational sports: mountains of waste produced by the city, covered with a proportion of its water supply in the form of snow. There would be no need to pre-freeze the water because the waste mountain range would soon become high enough to ski on.[8]

11. *Vertical Garden City in Datatown. This is an effective way of solving access to parkland and absorbing human carbon dioxide emissions.*
© MVRDV

Clone City exists in the present. It is repeated world-wide, from Los Angeles to Glenrothes, in the shape of a highway network linking housing estates, industrial and retail centres. It is, we are told, '[a] city formed by the mindless, market-driven proliferation of built environments', a place without a centre or a heart. Page and Glendinning argue that 'Cloning is . . . regimentation and anarchy at the same time' and hope for something better.[9]

They think one such option might be discerned in the most recent Scottish New Town, Irvine, described as 'large-scale new construction and zoned land-use, alongside a conservative regeneration of existing burghs and villages embedded within the pattern.'[10]

These criteria seem a little too tied in with an older, simpler view of city expansion. For one thing, the population densities come from another world than that of Datatown. While the Dutch work with population densities where 150+ people per acre (375/km^2) is a comfortable number and Datatown's 598 people per acre (1,477/km^2) is a realistic projection, the Scottish are concerned with 60–100 people per acre. In the end their analysis looks slightly out of date, even hidebound at times. The images they have chosen to illustrate reconciliation and 'the future' both show Victorian Scottish architecture and traditional religious practices. Their idea of the right kind of community seems to be rooted in same agreeable expectation as Matthew Arnold's in *Culture and Anarchy*: 'A certain number . . . of persons who are mainly led . . . by a general humane spirit, by the love of human perfection'.[11]

The forces shaping UK cities have been carefully analysed. One of the most thorough is the project examining 50 years' incremental growth of Swindon in Wiltshire.[12] The housing pressure in south-east England has led the Department of the Environment, Transport and the Regions (DETR) to declare a policy to build 4.4 million new houses by 2016. It is, to put it mildly, not popular and the planning framework has been used to resist local developments all over England. Resist – but not shape. The language of sustainability is being used everywhere, including Wiltshire, but the researchers note soberly that 'large-scale decentralised settlements might be more appropriate . . . than attempting to maintain strategies of containment'.[13]

'Managed counter-urbanisation' – construction of complete village-type communities in cities – is another option, as in Page and Glendinning's Irvine. It is the risk of failing utterly to 'manage' the forces of disaffected market-led expansion that is so unnerving. The MVRDV book summarises the essential questions. How much energy and from where? How large a water supply? How to clean the waste water? How much solid waste? Where should it go? Their book should be required reading for all policy-makers and planners.

Scotland, like everywhere else that is not frozen tundra or parched desert, needs an integrated vision of a whole urban landscape. If people cannot go and stay in the countryside, then nature must come to town. Planning should include the most recent technology for environmental preservation. The 'most recent technology' includes old-fashioned practices such as tree planting and mixed land-use. Dense but human-sized housing and neighbourhood trading should replace zoning, which wastes space and is boring. Local trade saves on transport, and provides local jobs and interest. A locality also means allotments and back gardens, parks and control of private cars in favour of public transport. Ecological construction is an important part of good management and incidentally a practical form of environmental education for the inhabitants. Criteria such as these are beginning to become familiar through area regeneration schemes, but rarely apply on anything like an effective scale.

There may be those who think all this amounts to no more than a paradoxical wish to return to an imagined past and also those who cannot see the point of the non-verbal, numerical cool of MVRDV. But 'Ecocity' is neither a throwback nor a Utopian dream. It is a way to help deal with an environmental crisis that is anticipated and debated all over the world. If Scotland does not join in, it will seem one of the most intellectually conservative nations in Europe.

The Ecology of Urban Communities

Alan Spence (among others) wrote about what old inner-city tenement Glasgow had to give to its people; it was a solidarity that transcended the physical awfulness.[14] The tenement idea is escaping the label 'awful' and is becoming seen as a means of civilised inner city living.[15] Its role in Ecocity is interesting because of the way it could be used for high-density, mixed occupancy dwellings close to the commercial hubs of cities, rather than the way it actually has been used either to pack in poor families or provide flats for the affluent few.

Cities have an intimacy and connectedness that is manifested in the tenement rather than the suburban semi or, even less so, in the results of *Clone City* house-building by construction firms intent on filling the green belt right up to the city boundaries. Contemporary gurus agree. In the sociologists' Bible *The Power of Identity*, Manuel Castells describes the city as a social powerhouse 'Urban social forces [have given rise to] an influential environmental movement, engaged around the world in collective survival.'[16] Will Hutton of *The State We're In* speculates in another context about trust and social sanctions in dense, city environments: 'But the construction of a stakeholder society is a physical as well as an economic one.'[17] He adds, after talking about the importance of close personal contacts in transactions: 'It is much better if these supply chains are ones in which there is regular ongoing physical contact. Now that's what I mean about urbanisation.'

What is the role of ecologically informed architecture and planning in this set of sociological and economic ideas? Not to sustain the tiny green construction industry would seem to miss a vital point.

Maybe the national sense of humour has been a reason why the British have been reluctant to accept the green message, with its overtones of earnest do-gooding. Still, by now there is some evidence that the English and the Welsh have taken it to heart. The grand plan for a millennium eco-village in Greenwich was a brilliant idea (the Dome may have drained it of vigour and cash). The London Design Museum has been holding annual, very successful competitions for environmentally sustainable architecture and design since 1999. In Birmingham a regeneration project has meant that a series of gentle pedestrian public spaces are rising from the concrete base of a flyover-and-underpass cityscape. Wales has got a wonderful new Botanical Gardens of Wales and Cornwall the extraordinary Buckminster Fuller-inspired bio-spheres. The Peabody Trust has provided several inspiring examples of good sites, as in their mixed-use, mixed tenure and high-density development of a former sewage works site in Sutton. The Sutton plans included 100 homes, almost 2,000 square metres of work space and lots of facilities such as playing fields, shops, clubs, a nursery and a health centre, all on about 3.5 acres of land.[18] Compared with this – and events move even faster in many other European countries – Scotland is slower and its city landscapes prove the point. The reasons are hard to understand.

Greening the Scotland's Built Environment

Scotland's built environment is under the control of remarkably few organisations. The national housing agency, Scottish Homes (set up in 1989 to manage social housing) had about £400 million to invest (£215 million from house sales, rental income and government grant income, and £160 million in private finance), but the figures have been reduced further now.[19] Deals with banks and other financial institutions supply more housing-directed investment, but have all the flaws inherent in the private finance initiative (PFI) idea. Private funding of course means that commercial considerations become paramount.

Local government was once – for better as well as, often, for worse – a big spender in the building sector. Local authority influence is now reduced to its powers as the planning authority. The potential clout that manipulating the planning regulations might have given local councils is counterbalanced by their anxieties about money and jobs. The talk is of partnerships and the purse-strings tend to be held by private finance and private construction firms. The latter do not just have the muscle to build. The current annual turnover of just one leading Scottish firm, Edinburgh-based Morrison Construction, tops £500 million. The building giants have a firm grip – or a stranglehold, depending on your point of view – on project size (large is better), land-use (green-field land preferable), transport provision (lots of roads and parking spaces are good) and building methods (as established with current suppliers and work-force training agencies, please). The increasing tendency to offer 'design and build' contracts, where the architect is a company man (or woman), has added to their pervasive power.

Of course there are other players, small but numerous and diverse, ranging from the state-supported housing associations, meant to provide some much-needed tenant influence, to tiny, fiercely independent groups like Community Self-build Scotland. There are determined brick structures built by women's collectives in Glasgow and self-reflecting abodes designed for wealthy clients in quiet corners of Perthshire. After more than twenty years in the doldrums, architecture as an art-form is becoming fashionable. The professional view of what Scotland should look like is a subject of real debate, stimulated by the new Parliament. September 1999 saw the publication of a consultation document on architectural policy.[20] The then Deputy Minister for Culture and Sport hoped that it would raise awareness and in the end lead to improvements in 'Scotland's architecture and Scotland's built environment'.[21]

The distinction between architecture and built environment was just about the most interesting thing said by the minister, but the suggestion that architecture should be seen as a contribution rather than an end in itself is not really elaborated in the beautifully produced document. Although there is a thoughtful section on the environmental value of architecture, the ambivalence comes out not only in the text, but also in the illustrations.[22] The text negates its own earlier evaluation by putting environmental sustainability into the category of 'Subjective, qualitative . . . social and cultural values', as opposed to the real stuff of 'Construction as process and building as product'.[23] The illustrations, with few exceptions, focus on buildings or details of buildings. Now and then, a fleeting impression of where it is placed is given by a view through an interestingly designed window or a cunning sight-line over a garden wall.

In May 1999 the architectural establishment, or at least the Manifesto group, produced a statement anticipating a ministerial policy proposal.[24] The manifesto was in fact about professional advancement. The item 'Face the World' turned out to argue for a Scottish Architects Export Agency and another, entitled 'Directing the Future', for an Architect General for Scotland, while 'Create and Live' was about setting up a Centre for Creative Architecture with plenty of Continuing Professional

Development courses – and so on. The contrast with the robust social and environmental arguments in the report by Richard Rogers's Urban Taskforce published almost simultaneously (June 1999), could not be more striking.[25] The report was commissioned by DETR for the specific purpose of guiding urban regeneration. It would have been good if Scottish architects had shown themselves to be alert to such issues. The report is a compendium of precise recommendations, which do not stop at buildings but treat area regeneration as a structural whole. Rogers developed his ideas about the 'Compact City' in the 1995 Reith lectures and in a book with the telling title *Cities for a Small Planet*.[26]

The Royal Incorporation of Architects in Scotland (RIAS) run an award scheme called The Regeneration of Scotland Award, but so far few of the awards have gone to green projects. The participants in the debate on 'Architecture is amoral' in the *Building Scotland* series agreed, overtly or tacitly, that environmental criteria on building probably amounted to unjustified constraints on creativity.[27] The journalist Penny Lewis, one of the panel members, singled out for criticism Scottish Homes's recent decision to demand energy-audited projects.

It is true that environmental good sense does not generate aesthetic pleasure as a by-product. There is no principle of natural justice causing less polluting and resource-hungry structures to look more attractive. But it is an absurd misunderstanding to assume, as many seem to, either that the reverse is true – good architecture must be unecological – or that eco-houses by definition must be boring. Green architecture has actually got a bad reputation. The architectural writer Jonathan Glancey declared 'Sustainable architecture [means] buildings that are bricky and crushingly dull. Boring Scandinavian houses covered with solar panels.'[28] It is true and it is not true. Generally, it should be easy to refute the supposition that anything eco looks dull: what are expensively trained architects for? Dragging great old buildings like the Parthenon into the argument ('lovely though un-green') is as specious as it is illogical. Worse is the suspicion that if you build for ordinary people – the people who are going to live in mass-produced housing estates, come what may – then all aspirations might as well be abandoned, aesthetic as well as environmental.

A group of people, mainly sociologists, geographers and planners with provocative views of what townscapes are about, met in Glasgow in 1996. The occasion was a conference called 'Images of the Street'. The street is a good subject for discussion: '[it is] located at an intersection of several academic disciplines'.[29] It is also the best vantage-point from which to understand a city. The reason why city natives are so often quite unaware of their surroundings ('I just live here') is precisely that they do not take the time to stroll along its streets and lanes and paths. But of course all the streets are important, because, unlike the visitor, the native often cannot choose his or her environment. Too many city fathers are pleased to use the picturesque or stately or ancient central scenery as stage flats, sheltering the drab areas where so many of the population lead their lives. The aspect that particularly interested the urbanists at the conference was the street and its extensions into squares and market-places as a safe meeting-place for the local community. Building for communities is also a central tenet for 'green' architects.

Scottish Architects with Green Tendencies

The so far rather gloomy view of Scotland's urban landscapes is not just based on reading and looking, but also on interviews with people in the know. Discussions with several eco-conscious architects from different practices have confirmed that their ideas have not been taken seriously, let alone prioritised. However, there is a growing understanding of both their approach and above all, of the advantages of green building technology. Below follow summaries of interviews with Howard Liddell, and his building engineer partner Sandy Halliday of the Edinburgh-based Gaia Group and with Fraser Middleton, who was an occasionally rebellious cog in the big machine of Fife Regional Council Housing Department (later, he left to work on his own). I gathered more interesting information and opinion from Colin Wishart, who works with Ian Dunsire (in Battledown

Studios: Natural Architecture, Perth) and from colleagues of John Gilbert, whose eco-oriented practice is based in the Templeton Business Centre, converted from a flamboyant carpet factory on the edge of Glasgow Green. Non-architects have also contributed greatly to my understanding of the process that shapes the built environment, and starts with land acquisition and financial deals and proceeds via often competitive planning stages to contracts, design and construction. Such contributors included John Gunn, a chartered surveyor and member of the Scottish Ecological Design Association (SEDA), Jim McFadden, a Glasgow Housing Association manager and Janet Brand, an environmental planning expert from Strathclyde University.

Interview with Howard Liddell and Sandy Halliday
(Gaia Group)

Finding the Gaia Group was a revelation. The atmosphere of dour endeavour that surrounds so much ecological enterprise seemed to lift and show a view of hope, enthusiasm and more than modest success.

Howard Liddell leads the Edinburgh group through a whirl of ceaseless activity with a rare combination of architectural imagination, theoretical insight and social commitment. He has many links to the world outside Scotland, notably Scandinavia (the group have an office in Norway), Germany and England. Still, perhaps the most admirable thing about Howard Liddell and his colleagues – notably Sandy Halliday, of whom more later – is that they make things happen. During the course of a year of being marginally involved with the team, I felt privileged to be allowed to join several interesting events.

One of the events I missed in the summer of 2000 was a workshop on Sustainable Rural Housing in Scotland, but by all accounts it was thought-provoking and useful.[30] Rural construction was one of its issues on which there is real contact with the Scottish Executive, where a group has been working on rural development policies for some time. Building in the countryside with local materials was a starting point for the Gaia practice, but its interests have now moved on to urban and industrial projects. Howard Liddell almost wearily reminded me of the success (awards, prizes, publicity) of the first Gaia 'eco-house' in Scotland, Tressour Wood (built in 1993 near Aberfeldy) with its timber construction, 'breathing' walls and organic paints. Other innovative rural buildings figure in their list of 'case histories', constructed using techniques such as roundpole frameworks, earth, straw, slab wood and cellulose and using new, low-energy methods such as dynamic insulation and pore ventilation.

The Gaia Group-led commission to construct a visitor centre in Glencoe for the National Trust is still under way. It is designed as a 'village' rather than a single building. The 'eco-village' is actually a working concept of sustainable living in small, more or less self-sufficient communities and something in which the group's taken a constructive interest. The National Trust commission was to some extent a 'top-down' project,

12. *Leslie Court in Fairfield, Perth – a recent addition (completed 1996), built to Scottish Homes cost criteria. The Fairfield Estate, which had become a slum, was refurbished and regenerated according to ecological principles and the Gaia Group's ideas about building.* © The Gaia Group

with the capital provided by a client organisation, but much of the Gaia Group's work has been driven by communities. This includes some of the larger constructions, notably community leisure and sports centres. Howard Liddell believes in community self-regeneration and has, characteristically, turned his belief both into a theoretical model and into practical projects. The model is a vividly designed flow-chart for interactions between *Agenda 21* (top-down) eco-policy and 'folk' (bottom-up) requirements and wishes.[31] The practical projects are many, and include the more than fifteen-year-old Fairfield Co-operative in Perth and more recent regeneration studies based in Renfrew and Paisley.

Getting support for projects depends above all on the client, but this cannot make up for the vagaries of the planning system and the sheer unavailability of technical supplies, from sustainably grown timber and heat-pumps to potentially local products like stone and bricks. When speaking with environmentally aware architects, these are the recurring themes. Howard Liddell commented on the pervasiveness of the networks set up by the powerful, conventional construction firms with their suppliers and established clients, but added that the local authorities are even harder to negotiate with. They have powers of planning and development that make them both key agents of, and major obstacles to change. With honourable exceptions, they tend to be reactive rather than proactive, and bound by existing practices and regulations, rather than prepared to use their influence in favour of innovation.

Operating in this climate is still frustrating, although it is almost ten years since Howard Liddell co-founded the Scottish Ecological Design Association (SEDA). SEDA is still going strong and has become widely recognised. It is providing evidence for the Scottish Executive and to interested gatherings of MSPs. An important part of SEDA is the series of events it organises. I have been present at a few of them, which was enough to make me enthusiastic about their vital role in providing access to technical information. One memorable speaker devoted his talk to the uncompromising-sounding subject Sustainable Urban Drainage Systems (SUDS) and made it fascinating. The idea is to find permeable alternatives to filthy and toxic water pouring from tarmac surfaces into the urban water system.

Technical understanding combined with real ability to communicate is rare, but the other key partner in the Gaia practice has got both skills. Sandy Halliday is a mechanical design engineer and an experienced specialist in ecological construction, who manages Gaia Research and co-manages SEDA (as its current Chairperson) and apparently effortlessly generates new ideas. She has written books with the kind of practical slant that are hard to find – less about speculative whys and wherefores, more about the how-tos, but her long publication list also includes research into sustainable building technologies.[32] Her organisational skills and genuine wish to get people talking about ideas often result in seminars and other get-togethers. Some are predictably technical and some are purely inspirational, like the seminar on animal architecture.[33] No, it was not about zoo cages or cattle-sheds, but about the structures that animals build. From mutually supporting frameworks in birds' nests to air-conditioning channels in termite mounds, the animals create shelters of instructive elegance and resource economy. Trust the Gaia Group to take more than a passing interest. It is not only investigating the engineering principles for human applications, but is also well on the way to designing a new display for Edinburgh Zoo.

Interview with Fraser Middleton

(then of Fife Council Architectural Services)

Fraser Middleton is a quiet man with a straightforward vision of decent housing. His knowledge of ecological building techniques seems to have come to him through his own interest in practical problem-solving, rather than being driven by theoretical speculation, political enthusiasm or dominant mentors. He was rare indeed as an environmentally aware Scottish local authority architect. Speaking to him after his decision to go it alone, he tactfully preferred not comment on the conditions of his local authority job. Still, it is not difficult to form a general opinion: council architect departments are becoming dull places, as new building commissions increasingly are handled elsewhere. It leaves – and that only for the time being – mostly just maintenance work to do.

When we first met, Fraser Middleton was still working within the local authority framework. The achievement he wanted to focus on was the construction of an estate with fifteen dwellings at Turner Crescent in the sprawling community of Methil. It is in a part of south-east Fife that has lost much of the old coal mining assurance and gained an unenviable reputation for disaffected young people. He had a realistic view of the expectations of people in this area and knew that fancy ideas would not be appreciated. The chance to design the estate from scratch was a rare one and depended on a complicated deal between Fife Council, which owned the site and were worried about their housing lists, and the local Kingdom Housing Association. There were more interested partners. East of Scotland Water (ESW) were consulted about the 'grey water' use and were keen to reduce the need for a clean water supply. Forward Scotland provided some funding and the Scottish Environment Protection Agency (SEPA) also took an interest in the project. In spite of all this officialdom, there were delays due to the fact that neither building regulations nor the legislative framework for construction contracts are supportive of ecological projects.

When talking to Fraser Middleton I felt encouraged to ask nuts-and-bolts questions. So what did he do? Well, he decided on a framework of relatively thin timber-struts, supporting an inner layer of plaster-board and an outer one of recycled particle-board, both without air-tight foil or plastic membranes, which are standard but cause condensation. The cavity insulation material was made from recycled paper-pulp. He had to fight for this, as he also fought for window frames made of laminated timber, which reduces saw-mill waste. They are ecologically preferable to the normally used PVC-type, but these have lower short-term maintenance costs. The amount of heat loss (through walls and roof) is an important factor and was intended to be comparable with the best European (Swedish) values of about 50–60 kWh/m^2. This is good, even if not as low a figure as the experimental German 'Passivhaus' at 15 kWh/m^2.[34] The Methil houses actually exceed the possible maximum achievable energy conservation level set by the National Planning Regulations. The gas-heating is boosted by solar panels and sun-traps, that is glazed porches with solid walls which serve sun-fuelled storage heaters. The asymmetrical house shape allows two floors for living spaces carefully angled at 15 degrees off-south (the garden side), while the single floor 'backs' with bathrooms and bedrooms are north-facing (the car park side).

Middleton faced the problem of sourcing acceptable materials locally and succeeded in some respects: natural linoleum from a long-established firm in nearby Kirkcaldy on the wood floor (not wood-block, to avoid glues) and rainwater barrels (for garden watering) donated by a large local distillers. The concrete blocks and the rough-casting came from Fife, but then it got harder. The timber window frames came from Denmark, the cavity insulation from Wales, the argon-filled triple-glazing from France and the natural clay (rather than cement) roof-tiles from England.

There were so many more thoughtful details, and there was so much careful costing going into these rather ordinary-looking houses, that it is difficult to imagine that the project could be criticised from any point of view, except possibly metropolitan aesthetics. Yet every inch of the plans seems to have been fought over, notably the cost and the 'pay-back' that local authorities are concerned to see. For instance, the grey-water plumbing that exercised ESW seems to have poor pay-back, because Scottish water is not metered, so does not have a cost.

I have not been back since January 1999, which was just a few months after the completion date. However, I talked to Fraser Middleton a year later, and the only known problem was that the solar cells are too efficient and provided more hot water than the residents can use. I hope the beech hedges (hedges are greener than fencing – literally) are filling out, and the meadow flowers are settling in. Why just a scattering of meadow flower seeds? Easy to guess: there was no money left for any planting.

There are other sides to built environments. Many architects have reservations about being principled and insisting on building according to one set of ideas, even those as sensible and appealing as ecological construction. There is the matter of getting commissions in the first place, which although less daunting than it became in the 1970s and '80s, is still a major obstacle to risk-taking. Then there are the frustrating problems of finding the 'right' materials and the financial backing to use them.

John Gunn told me about a European daily publication, devoted to the thousands of tendering opportunities for suppliers that 'the system' requires the building industry to announce on a Europe-wide market-place. It is a market with complex rules, many losers and few winners. For instance, the rule that at least six tenders must be invited for each source of supply means that of those who submit their proposals, the majority will have no chance of seeing a return for their efforts. The principled pursuit of green materials is being complicated by issues such as the 'embedded energy costs', for instance of materials that have been transported from far away. The European tendering rules only cease to apply for projects costing less E5 million (about £3.2 million), which is not a lot for the construction industry. Thus the stone for the new Scottish Parliament building in Edinburgh could not be specified as 'from Scottish quarries', because European sources must be treated on an equal footing. The only way of avoiding blocks of stone being hauled across long distances was for the architectural specifications to try to detail a type of granite only available in Scotland – clearly not a possible solution in most cases.

Another problem for ecologically principled architects and designers is that they work in an often deeply conservative world. Although to the outsider, these professionals can seem to be almost obsessed with new ideas, it is not easy to shift the emphasis of planning authorities, contractors and clients. Colin Wishart and his friend and former collaborator Geri Cruickshank, are both dedicated to aesthetic innovation, and made it clear how remarkably hard it can be to get any kind of 'odd-looking' design taken seriously. They were excited about the plan to house a cantilevered workshop-museum for outsider art (its proper name is the Scottish Centre for Art Extraordinary) within an abandoned harbour swing bridge in Leith. Sponsors were not interested, in spite of the social and green credentials of the idea, and an elegant model as persuasive as a work of art. Building on inner city brown-field sites is part of current conventional wisdom and pushed by grant-preferences from Scottish Homes (from June 2000) but it is likely to take a long time before another Cruickshank idea, interconnected living-pods elevated on steel legs, are seen stalking disused inner-city lots in Scotland.

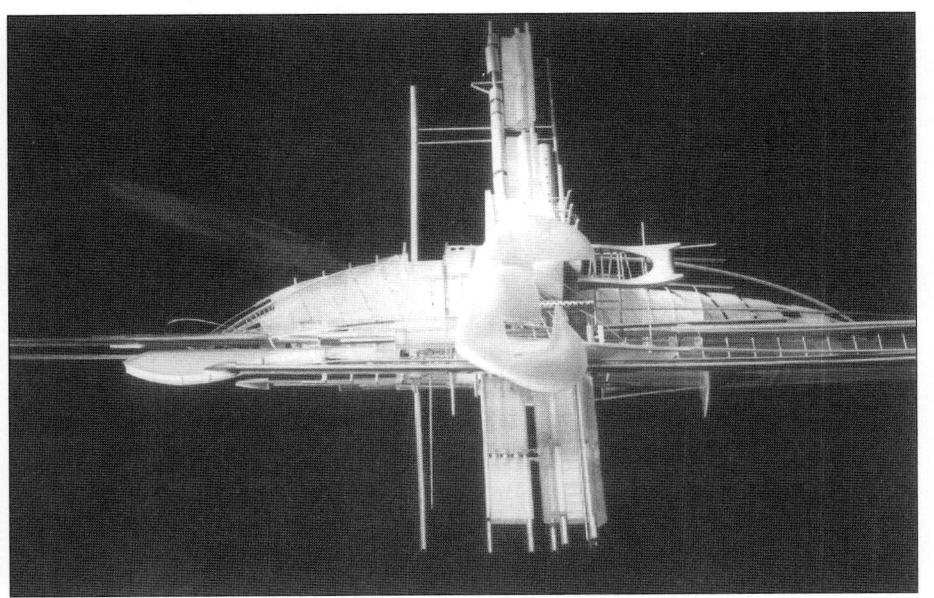

13. *A new concept in sustainable urban development: Geri Cruickshank's model for a centre for outsider art in a Leith brownfield site.*

The greening of the Scottish Planning Guidelines has been inspiring, but there is still much to be done. There are planners as well as architects who are exasperated by the process. Colin Wishart's example was the planning that has compromised so much of Dundee's western frontage along the River Tay. Both money and thought have clearly been spent, but without improving much on the fundamental incompatibility of a fine landscape setting and a jumble of commercial developments. The tourism and heritage industries have not managed to build anything that is more interesting than the big retailing structures. Once, the far-sighted City Architect Frank Thompson had a generous civic scheme for parkland, paths and walk-ways connecting the riverside with the city centre. However, the then Dundee District Council decided that development was about jobs and zoned the land for retail, with generous incentives. The plan's double failure is plain for all to see: one of the largest stores is empty, and the combination of roads and commercial developments have served to distance the city from its natural relationship with one of the finest landscapes of any city in Britain. True, it is not so bad as the meeting between the Firth of Forth and the edge of Edinburgh, but that is another story.

The Case of Glasgow: Glasgow and the Urban Environment

Glasgow has a central role in any argument about the urban environment in Scotland. There are few aspects of living in a city that are not vividly illustrated by the Glasgow experience, but the most talked-about are the most negative. Endless reports and plans have tried to solve the 'Glasgow problem'. Victorian reformers were appalled by its industrial slums. There were crucial improvements, including a good water supply from Loch Katrine, better housing and public transport. The next major step, the post-war (1943–6) Clyde Valley Regional Plan, insisted that 'co-ordinated administration and the improvement of the environment are essential to pull Scotland out of its recurring recessions'.[35]

In the beginning, these people-relocation exercises drew their structural inspiration from pre-war 'garden suburb' ideas. But places like 'Old' Pollok (largely completed in 1939) with its low-density cottage housing and undulating, tree-lined streets were too genteel to be functional. New towns, dormitory housing estates – 'schemes' – and frenetic road building took the place of Ebenezer Howard's marriage between town and country, which was to create a 'joyous union [from which] will spring a new hope, a new life, a new civilisation'.[36] Lack of time, tensions in the city council and tight finance resulted in municipal battles and abrupt decisions. But the urgency to build tall and fast came from the size of the population to be housed: 'Good enough houses in their day . . . but I mean, think of the block that could come out of there. You could get a hundred more families in that gap,' said a Glasgow builder with a social conscience, the grandfather in *Our Fathers*.[37] By the early 1980s, the bleakness of outer-rim Glasgow had become common knowledge. At the same time, the decay of the city core due to de-industrialisation (employment in Glasgow manufacturing almost halved between 1981 and 1991) and the growth of peripheral, car-dependent business parks and retail centres on the Clone City pattern all combined to make a dangerously deteriorating urban environment visible to the media.

Many reports based on commissioned research projects trace the shift from heavy industry to call-centre culture. The terrible health status and epidemic drug dependencies of people living in Glasgow are, surely, the results of deprivation and possibly a genetically static population (Glaswegians notoriously stay put). However, the class differences in living conditions, weak tax base and deteriorating housing seem related to a political paralysis that has overwhelmed even Scotland's new parliamentarians when faced with Glasgow. After another spate of distressing reports in the summer and autumn of 1999, the editorial in the *Sunday Herald* (5 December) started by asking: 'How much more sustained grievous bodily harm can the city take?'

The legacies of battles lost will haunt Glasgow for many years hence. No other Scottish town – not even Aberdeen, also torn apart by tough municipal management – bears so many scars. Glasgow has managed well: first, in changing its self-image, and then, some of the underlying socio-economic reality. All without losing its soul, as Ian Jack argued in a insightful essay from the mid-1980s.[38] But the impetus to change might peter out for lack of an effective and properly financed local government.

The high-minded planners of the 1930s and the immediate post-war period had ideas that still feel comfortable: all about gardens and allotments ('nature'), decent cottage housing, simple pleasures, family life. The significant differences between eco-oriented thinking then and now are related to changes in views on employment and family life. In Glasgow, the River Clyde and the East End do not provide jobs, and the 'Daddy-at-work-Mummy-at-home' social structure of fifty years ago has less relevance to how people lead their lives now.

One other striking change affects city spaces – streets, squares, markets – which used to be benign social places, but have become foci for a new 'fortress impulse'.[39] This defensive attitude and the guarded, policed precincts it has led to are discussed by Fyfe and Bannister in a fascinating summary of their work on city crime and CCTV. As a counterpoint, they refer to the praise of 'the sidewalk' and of tolerant diversity by entirely eco-acceptable Jane Jacobs. She has written both incisively and decisively about sources of vigour and decay in the city but pointed out that 'cities are immense laboratories of trial and error, failure and success in building and design.'[40]

In other words, the trick is to draw the right conclusions, without being dogmatic. What really went wrong in Glasgow, except for the unseemly haste with which the essentially benign planning principles were pursued? Apart from the familiar items, like lack of hard-hat employment, unstable family structures and growing rates of drug-dependency, the absence of local identity is a crucial factor. This is an elusive thing, not necessarily linked to prosperity, though it helps. Jane Jacobs is good on the subject of local identity, and insists that looking to factors such as income or physical amenities or aesthetic values 'is a waste of time'.[41] She analyses cases of planning provisions – parks, playgrounds, nice views, the lot – which apparently just accelerated the growth of slums and perceptively says: 'Our failures with city neighbourhoods are ultimately failures in local self-government.'[42]

Was it pity for a struggling population or admiration for the architectural interest of Glasgow that attracted the judges of the important competition for a British city to become the City of Architecture and Design 1999? Or was it an indistinctly articulated feeling that if the place was just spruced up a bit – given the design treatment – things would improve across the board? No one knows, but Glasgow won the competition. Quite right too, if only because a city with the wit to convert the blind-ending stump of overhead motorway into a comic-opera office complex, covered in shiny, ox blood-coloured tiles, deserves all the support it can get.

Still, for those of us who cannot stop wondering about identity, sustainability and the real world, Glasgow99 was not a cheering phenomenon. For all its stylishness and energy, it seemed to embody all that is wrong both with lack of sustainability and 'self-government' in the extended sense in which Jacobs uses the word. There were, I felt, too many failures to take people seriously and too much reliance on design-guru sensibilities rather than a realistic evaluation of environmental needs.

Glasgow Year of Architecture and Design 1999

Talking to the Director of Glasgow99, Deyan Sudjic, I got the impression that although he cares deeply about architecture, he distances himself from the earnestness of eco-building and community regeneration. Still, he disagreed when asked if the emphasis on green issues in the events and media-output of Glasgow99 had not been too weak, and reminded me of several punchy examples. For instance, there had been the Heisenberg Group and their curious creation of the Orange Journeymen (not Ulster marchers, but body-shapes made of orange plastic). These man-sized figures were put as temporary markers on city brownfield sites. There had been a charming, if a little cryptic, show of the plans by Norwegian school children for city regeneration and an exhibition of landscape designs. The main design exhibitions had included several ecologically conscientious pieces. But the landscaping was more about building than growing, and the eco-exhibits tended to get lost in the mass of un-green objects.

Then there was the idea to redesign decayed public spaces in Glasgow and construct new urban landscapes. It had real potential to change the city environment and was modelled on completed improvement plans in Barcelona. Originally it was intended to remodel fifteen city spaces. Only five were completed. The idea did not quite work out, since there was a lack – as Deyan Sudjic himself admitted – both of time to organise the projects and the funds to support them. I feel they were all to some extent over-designed and 'not useful', meaning that they had not been planned to include things like shelters, seats, pretty plantings, ponds, newspaper kiosks, play grounds, ballgame areas, cafés, market stalls or shopping arcades.

The concept of the Glasgow Partnership also did not quite work out, at least not in green terms. Individuals, organisations and communities had been asked to submit their own ideas and offered support with expertise and funds, when possible. Only a few could be classified as 'eco' by any standards – thirteen out of ninety-nine ideas, including the Journeymen and the landscape designs. The most celebrated ideas were visually arresting proposals, such as the multi-coloured illumination of the Cranhill water tower, described in the words of the *Working with Glasgow* booklet as 'a signal of the Partnership's Fund's achievements . . . blue, red and orange light from the grim grey tenements of Cranhill'.[43] To me, this project sums up much of what is worrying about current attitudes to social change: immediately appealing, easy and cheap without necessarily considering common resources.

Deyan Sudjic said that he had wanted to avoid 'evoking the nostalgia for the old Glasgow, covered in the sepia-shaded patina of old photographs'. He spoke of 'the city healing' that he felt had taken place during the year, which had brought physical changes but above all, mental ones. However, each physical project had meant complex deals and some disappointments, which sounds ominous for future change. The 'mental' projects were more successful but often marred by an almost desperate wish not to be serious. For instance, the large design exhibition in Kelvingrove Museum was meant to encourage public participation and concern about design, but few practical, technical or ideological points were made. Thrills and fashion seemed the only ways in which the curators felt they could involve people.

The blocks of flats at Glasgow Green could have been the crowning architectural achievement of Glasgow99. They are the outcomes of complicated negotiations with the city council, the construction industry, the sponsors, the architects and planners and many other interested parties. The fact that these stylish buildings are there and that the flats are being bought and lived in, is an achievement. But, still . . . I walked there, along tarmac paths across the Green, wondering at the lack of trees. The place was abandoned, apart from a group of men in orange safety vests investigating the damp grass for some unimaginable civic ill. There were too many cars circulating and the housing estates bordering the Green looked, uniformly and dully, like 'housing'. The Glasgow99 group of buildings was an island, almost eccentrically smart and unattainable-looking. The different units were constructed with little systematic regard for ecological building principles. An opportunity for innovation had been missed.

Green Building in Glasgow and Elsewhere

The Glasgow council housing stock is at the other end of the spectrum from the cool blocks of flats on the Green. The council has lost the struggle to maintain its decaying properties and building more council houses under the old system of finance is not an option. The political fights over the transfer of the entire stock to local housing associations (HAs) are all but over. It could be a very forward-looking move: local self-government would get an invaluable boost and the large council workforces, with their iron grip on labour costs, would be dispersed.

The HAs are nodal points in the complex interactions between the architects, planners, builders, consumers and the representatives of Scottish Homes, the Executive's housing quango. The HA is theoretically the client in these deals, although hemmed in by planning restrictions, cost concerns, grant regulations and marketing anxieties based on input from its members about 'what people want'. Often, the sheer number of considerations and the dislike of being seen to step out of line combine to make overworked people take fuzzy decisions.

The Glenalmond Street Estate, Shettleston

But in Shettleston, an inner Glasgow suburb, things had for once worked out nicely in 1998–9: An indefatigable HA project manager negotiated a Scottish Homes grant competition for a small housing development on the site of an old laundry at Glenalmond Street in Shettleston. The winning plan by John Gilbert had many features of sustainable construction and in addition, an innovative idea for the energy supply.[44] Flooding in the extensive mine workings under Shettleston has resulted in a huge reservoir of geothermally heated water. Although the water is barely tepid (about 12 degrees Centigrade), pumping it up into holding tanks and using heat transfer raises the temperature to about 55 degrees Centigrade. Some is then used to feed into the central heating, and some indirectly to heat the hot-water supply (to about 45 degrees Centigrade). Good insulation, together with further energy input from solar panels means that the estate annual heating bill is enviably low (in the order of £80 p.a. per dwelling). The upbeat forecasts of US experts in geothermal heat pumps (these do not generate but transfer heat) speak of huge reductions in overall running costs somewhere in the order of 75 per cent.[45]

The technology is imported and expensive. Only the US and Germany make heat pumps of the right kind. But then, only two Scottish firms (based in Dunblane and Forres) provide solar heating systems. Interestingly, one of them 'imports' solar panels from Wales and the other manufactures under a Dutch licence. By the way, it seems that Wales is becoming a very good source of high- and low-tech green products, for instance non-toxic, biodegradable insulation materials such as Warmcel (recycled newspapers).

On the other hand, some ingenious recycling was being done at Glenalmond Street, The bricks from the demolished laundry were not dumped (as building waste usually is), but used for building garden walls. Iron railings were found in a skip in Maryhill, but using recycled aluminium as a rust-proof alternative to roof zinc turned out to be impossible (not even the Fife-based branch of Alcan could help).

Dealing with the agencies involved was hard, as was confronting the people in the estate. It was a mix of rent-paying part-owners (a Scottish Homes idea), tenants and community care-sponsored individuals, and the HA swayed from side to side in the storms of opinion. There was a car-parking issue: the architects wanted gardens round the property, but some of the people wanted car parking. So did the local authority planners, who preferred 125 per cent, that is a quarter as many again as would be needed if every household in the block had just one car. In the end the street frontage looks a little sad: some car parking, some trees, tiny front enclosures (not really gardens) with mini-sheds hiding rubbish-bins. The back is almost entirely communal grass. People, the HA insisted, do not want gardens. It is too much trouble.

For the Gilbert Partnership, the constant need to revise the original specifications downwards runs through a politely phrased analysis of the Glenalmond Street project costs, written in collaboration with the construction company.[46] This, in spite of additional money from Forward Scotland, and the potential value of using 'Scotland's sleeping lagoons' to generate heat. The phrase was the headline in a small article in a local newspaper about the next HA to engage John Gilbert's team for building a housing scheme in Lumphinnans, a Fife mining area.

The Housing Association

The HA principle has by now become very popular – there are well over twenty in Glasgow – and it seems to appeal in particular to politicians. It is indeed an attractive idea, which encourages people to look after themselves and take responsibility for their own community housing. But these desirable things are not yet dominating the picture. Instead many HAs seem to be subjected to conflicting directive forces.

Different factors determine the size of the HA Grant (yes, the HAG) from Scottish Homes in support of the HA's building activities. For any one project, the HAG is proportional to the calculated rental income and can cover as much as 80 per cent of costs. The HA must raise 20 per cent, either as loans or by drawing on surplus rental income. The formula becomes complex when it takes account of things such as maintenance costs – there is a fixed allowance which does not cover major repairs – depreciation or higher than normal costs, because of special circumstances. The whole building exercise is monitored by Scottish Homes and managed on the spot by a Development Officer, who is a co-ordinator for everything, from finding plans, to land acquisition, to construction contracts, to property management.

Regenerating the Built Landscape

At the end of much research, for me it is planning which remains the least understood area. Even after speaking at length to an academic planner, I felt no clearer about how this crucial system is operated. Who decides how great a housing density, say, before the plans must include a health-centre, a shopping area, a public transport provision, work places, parks? New building and regeneration developments alike depend on the planners, but apart from guidance notes, the state planning system seems based on the elusive principle of *ad hoc*.

'Urban regeneration' has been a mantra intoned by politicians, architects, planners, social reformers and practically everyone else for many, many years now. David Page, who is one of Scotland's leading architects, summed up the end of the 20th century: 'What follows [the 1920s and '30s] are three decades of stigmatised Black Hole – followed in turn by the Disenchantment of the 70s and the resulting attempt at repair, renewal and reconsideration of the city, lasting to the present.'[47] He felt that architectural development in Scotland had largely missed the late-20th-century boat, but paid homage to the innovative design of some 'carefully thought-out small buildings', genuinely new but one-off creations. Page has argued more than once that 'tenamental repair' tends to be part of the blight aspect of urban regeneration, of 'disenchantment' rather than 'enlightenment'.[48] If one substitutes 'brownfield site building' for 'tenamental repair', this line of thought becomes connected up to the current agonised debate about designing the new versus refurbishing the old, which is focused on the heavily built-up landscapes of south-east England.

Not only the architects but also their direct or indirect clients – the citizens at large – would surely prefer to have new buildings in their own image. Regeneration, from that point of view is regressive, the mend-and-make-do consequence of derivative committee solutions to the problems of a timid and overpopulated country. But that last, simple fact – the British Isles are overpopulated – imposes restrictions that the more complex, social arguments in favour of regeneration might have failed to enforce. As the landscape is increasingly redesigned to serve urban populations beyond the limits that traditional, land-intensive farming pursuits allowed, land is becoming almost too precious a resource to cover with new buildings. One option is urban regeneration on a grand scale. Another is new building that is ecologically state-of-the-art and designed discreetly to integrate into existing landscapes or cityscapes.

The problem with make-overs of whole cityscapes is that each project becomes a bureaucratic and financial obstacle-course. The Glasgow99 experience is only one example of how near-impossible it is to parachute in and 'renew and revitalise'. Even the attempt to create a smart traffic-controlled shopping area in central Glasgow, backed by serious funding and the might of Glasgow City Council, is beginning to get frayed round the edges. The reason this time is not so much the scale of the original project, but of paying for the people and the skills required to maintain it in good shape, in the face of constant urban wear and tear.

Really successful attempts at regeneration tend to be limited, well funded and backed by some continuing commercial interest. The outstanding Scottish example is probably New Lanark, one of the few Scottish entries in the list of schemes picked by the British Urban Regeneration Association (BURA), an independent ('arms-length-from-government') organisation supported by the great and the good.[49] New Lanark is a model from many points of view. In the first instance, Robert Owen's late 18th-century version of an imaginative and high-principled rural development project was protected by enlightened and properly enforced planning decisions 200 years later. By the early 1970s, the old community centred on cotton-mills had become a depopulated industrial slum. An effective partnership grew round the regeneration project and now this fine heritage site pulls in about 400,000 visitors annually, has got a permanent population, add-on businesses and permanent jobs.[50]

14. *New Lanark industrial heritage village in Lanarkshire, showing renovated houses and mill buildings.* © Colin McPherson

Even the recreation of living, happy New Lanark took decades of effort, in spite of its cluster of A-listed houses and mills, an idyllic setting and interesting history. The result seems set to last though, unlike so many of the sprucing-up jobs done in less promising built-up areas. Regeneration of the deep countryside is harder still, unlike the built-up areas that can be turned round at a relatively fast pace. It might, however, be the countryside that becomes the source of real and meaningful change in the construction industry. One day, which cannot come too soon, the National Planning Guidelines must start pushing the eco-building revolution in earnest. Then local resource retailing, effective recycling enterprises and community construction firms might spring into life and help to support a truly sustainable countryside economy.

But the cityscapes too must change to become not only more humane in scale and operation, but also more sustainable in terms of resource-use and more closely allied to the land, the existence of which so many urbanites tend to forget. The cityscape should reflect the landscape, not obliterate it. Better a burn with footbridges than a flow of water in an underground culvert; better steeply sloping up-and-down streets than tunnels and viaducts – and so on. One reason that so much about modern cities ends up failing to enchant or even satisfy, is that the marriage of land and built environment does not work out in a way that meets human needs, of which beauty is only one. Surviving fragments of old cities are often described as showing the characteristics of 'organic growth'. Intuitively, we understand this to mean that some past magic led to the fusion of beauty and utility, of great buildings and idiosyncratic but liveable-in houses. Are we unable to create 'organic growth' in that sense now?

A rhetorical question. Here are some suggestions that may hold at least partial answers from a favourite source of ideas, a publication called *Trespassers* compiled from the now mysteriously defunct Netherlands Design Institute.[51]

First, The Sponge Company, run by the 'space broker' architects J. Vink and R. Heijne. The company name reminds you that there is a species of quiet multi-cellular beings, whose very existence teaches that you can be both humble and creative about use of space.[52] J. Vink says:

> The Sponge Company fills a gap in the market . . . We make money by intensifying the use of city space . . . Cities are like malfunctioning sponges. They too consist of systems of interlinked cavities, but somehow their ability to use them to the full is inadequate . . . There are wastelands, whole buildings . . . that suffer from long-term emptiness or only fulfil some sort of part-time function.

Second, the satisfying concept of the 'architectural laziness', launched by architect Willem Jan Neutelings, who says:[53]

> Ambitions combined with diligence is extremely dangerous, but ambition with laziness provides a neat balance, because it demands creative solutions

The editors comment:

> [For example] in the middle of this rather austere modern complex, he integrated an existing old residence with its garden . . . Before the intervention, or rather, absence of it, there was just your unnoticed old villa, but now it has gone through a strange process of revival.

Notes

1. *A Century of the Scottish People 1830–1950* by T. C. Smout, 1997, p. 33.
2. *The Gaia Atlas of Cities. New Directions for Sustainable Urban Living* by Herbert Girardet, 1996; 'Green Cities' by Vanessa Baird, in 'Green Cities. Survival Guide for an Urban Future', *The New Internationalist* 313, June 1999, pp. 7–11.
3. *The Habitat Agenda* is the official document of the 1996 UN Conference on Human Settlements (Habitat II). UN website: www.unhabitat.org.
4. *Metacity Datatown* by MVRDV, 1999. Cf. www.archined.nl/010
5. *Area Regeneration. Social Inclusion Research Bulletin* 1, The Scottish Office Central Research Unit, March 1999 Cf. www.scotland.gov.uk/cru/resfinds/sir01-00
6. Ibid. Project funded by the Economic and Social Research Council.
7. *Clone City. Crisis and Renewal in Contemporary Scottish Architecture* by Miles Glendinning and David Page, 1999.
8. *Metacity Datatown* by MVRDV, 1999, pp. 190–5.
9. *Clone City. Crisis and Renewal in Contemporary Scottish Architecture* by Miles Glendinning and David Page, 1999, p. 5.
10. Ibid., p. 181.
11. *Culture and Anarchy* by Matthew Arnold, 1882, p. 85.
12. *City for the 21st Century? Globalisation, Planning and Urban Change in Contemporary Britain* by Martin Boddy et al., 1997.
13. Ibid., pp. 329–30.
14. *Its Colours They Are Fine* by Alan Spence, 1983.
15. 'Tenements on Trial' in the series of 'Building Scotland' discussions, led by Isi Metzstein, David Page and Fionn Stevenson. Organised by the Royal Fine Art Commission for Scotland, Edinburgh, August 2000.
16. 'The Power of Identity' by Manuel Castells, in *The Information Age: Economy, Society and Culture*, 1997, p. 62. See also pp. 110–33.

17. 'Urban Regeneration, Architecture and the Stakeholder Society' by Will Hutton. Debate in *City*, 7, May 1997, pp. 67–70.
18. The BedZED Project, developed by the Peabody Trust with environmental specialists BioRegional and architect Bill Dunster. Personal communication (press release material). The Peabody Trust, January 2000.
19. www.scot-homes.gov.uk/ The data in the text are from 1999–2000. Scottish Homes will invest £215 million in its housing development programme in 2000–2001.
20. *The Development of a Policy on Architecture for Scotland*. A consultation document from the Chief Architect's Office. The Scottish Executive, September 1999.
21. Ibid., p. 1.
22. Ibid., 'The Environmental Value of Architecture', p. 15.
23. Ibid., 'The Process of Building', p. 22.
24. 'A Manifesto for an Architecture Policy for Scotland', in *The Case for Change: an Architecture Policy for Scotland* by Peter Wilson, 1999, pp. 13–15.
25. *Towards an Urban Renaissance*. Report by the Urban Taskforce led by Lord Rogers of Riverside, to the UK Department of the Environment, Transport & the Regions, 1999.
26. *Cities for a Small Planet* by Richard Rogers and Philip Gumuchdjian, 1997.
27. 'Architecture is Amoral'. Debate in the series of 'Building Scotland' discussions, led by Peter Jones, Penny Lewis and Dorian Wizniewskiavid. Organised by the Royal Fine Art Commission for Scotland, Edinburgh, August 2000.
28. 'A Thing of Beauty is a Joy for a Very Long Time' by Jonathan Glancey. *The Guardian*, 5 July 1999.
29. *Images of the Street: Planning, Identity and Control in Public Space* by Nicholas R. Fyfe (ed.), 1998.
30. *Sustainable Rural Housing in Scotland*. A Workshop Review. Gaia Group, 2000.
31. *Sustainable Community Development Model* by Howard Liddell, 1999.
32. *Green Guide to the Architect's Job Book* by Sandy Halliday, 2000.
33. 'Animal Architecture'. Proceedings of a Seminar at the Edinburgh Zoo. Gaia Group and the Department of the Environment, Transport and the Regions, 2000.
34. *Natural Capitalism. The Next Industrial Revolution* by Paul Hawken et al., 1999, e.g. p. 103.
35. *Grieve on Geddes* by Robert Grieve, 1980, p. 11.
36. Introduction: 'The Matter of Englishness', in *Regaining Paradise. Englishness and the Early Garden City Movement* by Standish Meacham, 1999, pp. 1–10.
37. *Our Fathers* by Andrew O'Hagan, 1999, p. 154.
38. 'The Repackaging of Glasgow' by Ian Jack, in *Before the Oil Ran Out. Britain in the Brutal Years*, 1997, pp. 200–20.

39. 'Changing Places: Perspectives on the Development of a Municipal Suburban Landscape' by Gerry Mooney, pp. 31–43 and '"The Eyes upon the Street". Closed Circuit Television Surveillance and the City' by Nicholas R. Fyfe and Jon Bannister, pp. 236–53, both in *Images of the Street: Planning, Identity and Control in Public Space* by Nicholas R. Fyfe (ed.), 1998.
40. *The Death and Life of Great American Cities: The Future of Town Planning* by Jane Jacobs, 1961, p. 16.
41. Ibid., 'Slumming and Unslumming', pp. 284–304.
42. Ibid., 'Governing and Planning Districts', pp. 418–42.
43. *In Partnership. Working with Glasgow for 1999*. UK City of Architecture and Design in Glasgow, 1999, pp. 8–11.
44. *Scotland the Brave. Innovation in Housing*, A review of selected projects by John Gilbert Architects, 1999, pp. 13–16.
45. Heat Pumps. Case Studies and information, see e.g. US Department of Energy, Geothermal Energy Technical Website, at http://id.inel.gov/geothermal/heatpumps (3[a]) another interesting site is www.geoexchange.org/cases/dircases.htm
46. 'Cost Study: Shettleston Sustainable Housing' by John Gilbert and Armour Construction Consultants. *Building*, 6 August 1999.
47. 'The Censorship of Neglect: Morris & Steedman's Private Houses' by David Page, in: *Rebuilding Scotland: The Postwar Vision 1945–1975* by Miles Glendinning (ed.), 1997, pp. 115–25.
48. 'Tenements on Trial' in the series of 'Building Scotland' discussions, led by Isi Metzstein, David Page and Fionn Stevenson, organised by the Royal Fine Art Commission for Scotland, Edinburgh, August 2000.
49. British Urban Regeneration Association www2.rudi.net/ppo/bura/award97
50. The partners are: the New Lanark Conservation Trust, including the New Lanark Association, which is a registered Housing Association, the local authority, the Local Enterprise Company, Historic Scotland, Scottish Homes, the European Union, charitable trusts and various sponsors from the private sector. Information, e.g. www.newlanark.org/education/index
51. *Trespassers: Inspirations for Eco-efficient Design* by Ed van Hinte and Conny Bakker, 1999.
52. Ibid., pp. 20–5.
53. Ibid., p. 16.

8

The Environmental Economy: Research and Development, Local Trade and Big Business

The Economics of the Environment

SPECULATIONS ABOUT WHAT a 'greening' of the First World states would look like always start with reductions in consumption. In *Scotland: the Challenge of Devolution*, Kevin Dunion quotes a raft of figures illustrating the gap between the OECD members and the countries that belong to the 'world's low income group'.[1,2] There is about the same number of people in the low as in the high camp (respectively 19 per cent and 21 per cent of the world's population), but the figures for OECD resource consumption are not just much higher, they are grossly so. The First World takes 83 per cent of the global GDP, uses 75 per cent of its energy production and owns 92 per cent of its private cars.[3] The OECD policy statements bristle with the phraseology of sustainability, and the organisation aligns itself with other international conglomerates in pushing for cleaner and less wasteful technologies, with the inevitable asides about Best Practicable Environmental Options (BPEOs), that is, essentially, cost versus benefit calculations.

Anyway, effecting change on the scale required is dauntingly problematic. Radical changes are facing all other capitalist, free-market economies over the next 20–50 years. To paraphrase Kevin Dunion again, the challenge for the Scottish Executive is to set in motion policies that ensure, for instance, that by the middle of this century CO_2 emissions have fallen by almost 80 per cent, and consumption of non-renewable materials (e.g.

iron, aluminium, cement) by 85–90 per cent.[4] Can it be done, while holding on to the democratic and free trade ideals no one wants to abandon? Answers have included Utopian fantasies, but also the 'sensible capitalism' of John Elkington's *The Green Capitalists* and Amory Lovins's persuasively practical scenarios managed by an enlightened technocracy.[5, 6] Lovin appeals to the fixer in us all and his bouncy predictions are appealing: 'Our thesis [is] that 90–95 per cent reductions in material and energy are possible in developed nations without diminishing the quantity or quality of services that people want. Sometimes such a large saving can come from a single conceptual or technological leap . . . more often, from systematically combining a series of successive savings.'[7] The idea is to create a new economy, 'Capitalism as if Living Systems Mattered'. Elkington, though less free-wheeling, held similar beliefs and was driven by the same optimism: 'Perhaps we are seeing the emergence of a new age capitalism, appropriate to the new millennium, in which the boundary between corporate and human values is beginning to dissolve.'[8]

Both prudence and a sense of justice are inherent in the concern about global resource consumption. As a context for supporting the development of new technologies, the up-beat business approach is much more attractive than just downscaling, for innovative products can be sold and generate wealth. The debate has moved from reducing consumption in isolation to setting environmental goals for R & D.

The general feeling among people in the know seems to be that although the Scottish establishment has woken up to these issues, it is neither familiar enough with the arguments of sustainable economics nor willing to act decisively. In their useful but mainstream guide to the Scottish economy, Jeremy Peat and Stephen Boyle approach the economics of the environment in their own hard-nosed way.[9] The asset that is Scotland's landscape is described more than once in the familiar postcard-view terms: 'Rugged beauty . . . space and grandeur . . . cultural heritage'. The environment is important enough for the reminder that 'Scotland must strive not to lose this environmental advantage.'[10] It provides a basis for a quality lifestyle that attracts inward investment managers and helps reverse outward migration, especially by candidates for the brain-drain. On the other hand its charms seem less vital for the tourist industry, which should worry mainly about the stagnant level of service provision and the strong pound. As for retail business developments, here environmental approaches affecting, say, overheads or car-use, are thought to be positively disadvantageous: 'Good environmental intentions may have . . . adverse effects . . . in damaging competitiveness.'[11]

This kind of analysis is widely accepted, and so clearly is the guide's complete omission of environmental technology-based industry and manufacturing. In order to change the economic evaluation of Scotland's landscape from 'nice backdrop' into a more dynamic role in business development, there must be legal and economic instruments that are deliberately designed to change the market geography.

To illustrate the point that pricing environments can function as the first step in an economic process, a recent summary of possible Scottish applications of such ideas listed: timber and non-timber benefits of 'multi-use forests', stock-management of red deer (too many) and wild salmon (too few), charging for recreational use of the countryside and agri-environmental policies.[12] This is taking the 'supply side' of environment – the landscape itself – seriously. It is a way of thinking that has got many and intriguing implications. Academic periodicals on the subject are beginning to emerge, which is always a good guide to the intellectual climate. These have titles such as *Journal of Environmental Planning and Management* and *Environment and Planning* and contain articles with jolly titles such as 'Environmental Management: Testing the Win-Win Model' and more sober ones, such as 'Environmental Performance Evaluation in the Water Industry of England and Wales'.

The list of the four Scottish examples of 'land use interventions' led by new economic instruments included only options for rural industries. No surprise there, because 'the environment' tends still to be very much associated with 'the countryside'. However, there are no hard-and-fast distinctions between what kinds of measures would work in countrified contexts and in urban ones. In all settings the most critical process, initiated by fiscal or other means, is research and development and design (R, D & D). Although the value of working on R, D & D in new green technologies is widely recognised, in Scotland few are actually doing it and those who try to seem to find it very difficult to get financial support.

Research and Development (R&D)

It is widely accepted that too little of research, design or development are being done in Scotland, not only with regard to environmental goals, but across the board. There is no argument about the fact that 'the level of business R&D is notably low as compared with the rest of UK' and in per cent of GDP less than half of the UK figure.[13] I can find no information about specifically environmental business R&D, but suspect it is minuscule in spite of its potential importance.

This cannot be for want of rhetoric. Echoing the DETR line, a previous Scottish Minister for the Environment (and Transport) said among other things: 'Business has to see sustainable development as an opportunity for new ventures and moving into emerging technologies.'[14] The next Scottish Minister for the Environment (and Sport and Culture), who was much more cautious on the subject, worried away about environmental indicators – 'Forty may be too many for ordinary businesses' – but spoke proudly of the Executive's record on 'investment in renewable energy'.[15] Forward Scotland with its remit of 'promoting sustainable development in Scotland through innovative practical projects' actually had no serious schemes for investing in new technologies in 1999, but its next annual report included 'Developing environmental service businesses in the Highlands and Islands' (that is, collecting waste) and a vermiculture (that's worms producing compost) project.[16]

But it would be unfair to expect a small quango like Forward Scotland to broker the kind of change that is needed. It manages a number of revenue sources for small project grants, including some New Opportunities funding and, especially, the almost too cunning Landfill Tax Credit Scheme.[17] Scottish Enterprise, a much bigger player, joined forces with one of Scotland's more environmentally oriented universities, Heriot-Watt in Edinburgh, and in 1999 set up the Scottish Institute for Sustainable Technology (SISTech).[18] It should be good, because the staff is working on 'improving links between business and research institutions ... identifying needs and opportunities for R&D in sustainable technologies ... securing funding, managing research ... developing business opportunities' and much more in the same vein. It is located in the countryside (the Tweed Horizons Centre in the Borders), but seems not to have the links to industry or other tertiary education centres that it needs. The staff is mysteriously tiny, that is, half-a-dozen graduate assistants and a director, who is actually the Professor of Civil Engineering at Heriot-Watt and besides a member of the Board at East of Scotland Water. Also, the new institute has few projects actually under way according to a personal communication in response to an inquiry.[18]

Heriot-Watt has a Centre for Environmental Resource Management, set up in 1996 as the result of a merger with the Institute of Offshore Engineering and Orkney-based International Centre for Island Technology. The centre draws on environmental expertise at Heriot-Watt University, which '... has always emphasised the multi-disciplinary nature of environmental and technological research and the critical importance of linking these research programmes both inwardly to the teaching programme, and outwardly to the needs of government, industry and the wider community'.[19] It runs some environmentally credible-looking postgraduate courses. However, I can find no mention of SISTech at all. The Centre for Island Technology has got a link to the virtual University of the Highlands and Islands, which will be discussed later (Chapter 9).

Meanwhile, there is environmental R&D going on elsewhere in Scotland. Systematic programmes are generated by the Environment Group, now part of the Scottish Executive's Rural Affairs Department (SERAD). Most of the group's projects target waste management and pollution monitoring, or countryside and natural heritage issues, as in 'Levels of Volatile Organic

Compounds (VOCs) around Whisky Distilleries and Bonds' or 'Natura 2000 Socio-Economic Audit'. The rare R&D-type projects are vaguer, for example 'Developing Markets for Recyclable Materials' (make materials that are easy to recycle?) or 'Scottish Houses: New Build Home/Work Units in Rural Scotland'[20]

The Strategy for Agricultural, Biological and Related Research for 1999–2003 should be important for rural eco-industries, involving as it does a whole group of prestigious institutes and university departments, some of which have made policy-statements about sustainable goals.[21] The research briefs are wide-ranging, including food product quality, animal welfare, organic farming practices and land use optimisation. Outside this sphere, though, it is hard to find systematic research programmes on sustainable technology.

What is done is nowhere near enough. Comparisons with 'abroad' are humbling, whether one goes to the huge and burgeoning eco-market-place of the United States, or to the smaller European equivalents in – especially – Germany, The Netherlands and the Scandinavian countries. It is impossible to survey even a fraction of what is going on, without writing another book or two, so a few examples will have to suffice.

At the Rocky Mountain Institute, the shrine of Factors Four and Ten (that is multiply your output, and your profits by four or ten, while cutting down resource use), the Lovins and their colleagues practise the resource-saving ingenuity that they preach. The institute itself is a lesson in smart, energy-conserving building technology. It is involved in endless practical projects, from green planning to reducing car-dependence in a California town to boosting worker performance and till takings by 'sophisticated day-lighting' in Val-Mart's Eco-Store in Kansas. Specific fixes come in at all levels – cutting water use in homes and industries by automatic leak-monitoring equipment and used-water cleaning- and recycling systems; super-efficient paper- and wood recycling plants and so on – but increasingly, their emphasis has shifted to a debate about policy-systems, including rational valuation of 'natural capital'.[22]

This too is a concern of the European Roundtable on Cleaner

Production (ECRP), one of many European organisations in this area. There is a whole international bureaucratic framework for regulating environmental management systems and audit, waste minimisation and cleaner production. Firms get a 'clean' certificate after fulfilling the criteria specified by the accreditation bodies for registration as either meeting an international standard or its European version (the ISO 14000 family and relations) or the EU Eco-Management and Audit Scheme (EMAS).[23] The EMAS regulations were adopted by the Council of the European Union and the European Parliament in February 2001. In view of this, firms which cannot add EMAS or an eco-standard certificate to their product specifications should start really worrying.

15. *Windfarm in Datatown. The MVRDV Group calculated the extent of the windgenerator capacity (760 m height, in a zone of 10 x 400 km) needed to supply 78,000 MW per year. (The Netherlands uses 11,000 MW per year.)* © MVRDV

In 2001, the 7th ERCP meeting selected 'Sustainable Production and Consumption Systems – Co-operation for Change' as its main theme. It proceeded to address about thirty sub-themes, all related to eco-efficiency and regional sustainable development. The lead institution was the International Institute for Industrial Environmental Economics (IIIEE) at Lund University in Sweden. To what extent does it matter for a small, geographically marginal and industrially vulnerable country to be into sustainable technology to this extent? Scandinavian countries do care more than most, and have done for many decades. The main concern now is the possibility of losing leadership to countries such as Austria and Canada, where huge investments in terms of percentage of GDP have been made in order to get into a world market with an estimated value of some £400 billion.[24] The investment process was kick-started in the 1970s, when relatively stringent environmental legislation was enacted in Scandinavia. As a recent study showed, Norway, Denmark, Sweden and Finland are 'best in the world' when it comes to the Life Cycle Analysis (LCA) methodology. LCA means examining products – cars, electrical equipment, pulp and paper, construction materials and buildings, chemicals etc. – from the point of view of sustainability, from manufacture to ultimate disposal.[25]

There are more concrete examples. What about Denmark's outstanding success in exploiting the market for renewable energy generation? The box illustrates the point: it contains a list of the areas in which Folkecentret (The People's Centre for Renewable Energy) in Copenhagen is active:[26]

Biomass, Biofuels and Energy from Waste Applications; Biogas plants; Biomass, Biofuels and Energy from Waste Applications; Biomass, Biofuels and Energy from Waste Services Consultancy; Biomass, Biofuels and Energy from Waste Services Design; Biomass, Biofuels and Energy from Waste Services Training; Photovoltaics Applications Hybrid Systems (solar and wind); Photovoltaics Applications – Integration into Buildings; Photovoltaics Applications – Rural Electrification; Photovoltaics Services Design of

Systems; Photovoltaics Services Training; Wave and Tidal Energy Services Design; Wave and Tidal Energy Services Training; Wind Energy Applications Hybrid Systems (solar and wind); Wind Energy Applications; Stand-alone Systems; Wind Energy Components Generators; Wind Energy Services Consultancy; Wind Energy Services Design; Wind Energy Services Monitoring; Wind Energy Services Testing and Services Training

From recycling cars (the EU is wrangling about the finer points of its Directives at the time of writing) to the market for sophisticated, low-impact public transport vehicles, from producer responsibility for recovering electric goods and 'white-ware' to food labelling, the market-place is both complex and expanding. It is not an easy one to make money in either: the demands for novelty, and the problems of expensive producer commitments and take-over-hungry competitors are as great or greater when compared with 'normal' products. Lack of experience and lack of national backing can be fatal.

Interview with Dr Thomas Lindhquist
(Assistant Professor and Deputy Head of the
International Institute for Industrial Environmental Economics
(IIIEE) in Lund, Sweden

Thomas Lindhquist, who is the supervisor and mentor for many of the students at this exclusively postgraduate institute, started out with a mixed bag of qualifications in civil engineering, economics and languages, including Russian. Recruited by the Lund University's forerunner to IIIEE to work on a new idea in environmental production technology – proactive environmental protection – he became fascinated by this 'interphase' subject area. He has been published on links between the 'producer-takes-responsibility' concept and recycling, and has also worked with firms to set up 'proactive' manufacturing schedules. By the late 1980s these ideas were well established and Thomas Lindhquist gained international recognition for a variant concept called Extended Producer Responsibility (EPR) and audit criteria, the 'Environmental Descriptors', which became part of the product's Life Cycle Analysis (LCA).

These ideas, derived from practical experience but developed into generally applicable principles for 'eco-manufacturing' and marketing, were taken up with some enthusiasm by the then Director of the Swedish Environmental Protection Agency (NVV) and later also by the Swedish Department of the Environment. NVV was able to initiate effective legislation quickly. It seemed attractive to extend the producers' responsibility into manufacturing with an eye to post-sale waste management, precisely the point of EPR. This argument was linked to recovering the price of recycling (by the producer) in the price of the goods, that is the more wasteful, hard to recycle etc., the more expensive.

By the mid-90s Lund University was being funded by the Swedish state to create a partial replacement for its previous institute and IIIEE was the result. It is now earning considerable outside funding, notably from Europe but also from tendered research projects and industrial sponsorships. Sweden has got other institutions that, like IIIEE, focus on technical, economic and legislative aspects, but they are less common than research and teaching institutions that have their environmental base in science and engineering subjects.

What are the Lund institute's specialisations, compared with the Stockholm Environment Institute? It too is an example of a generalist institution with staff concerned with practical implications of environmental good practice on an international scale. The Stockholm Institute is funded directly by the state and is not linked to education. Its briefs usually involve getting teams of technical specialists together for the purpose of studying and preferably solving tangible, on-the-ground problems over a wide range: impacts of large building projects, water and sanitation in cities, forest management and so on. IIIEE focuses on working with companies in the market-place and on managerial or political environmental analysis. It has big international projects under way, notably in India and in Eastern Europe, where the institute has branches in Kalingrad, Smolensk, St Petersburg and Sofia.

I asked Thomas Lindhquist to speculate on differences between countries with respect to how they manage environmental issues. One of his comments was that the United Kingdom tends to prefer green activity, that is linked to nature conservation and ecology, while the Scandinavian countries and Germany, also highly eco-conscious, have invested much more in enterprise and legal requirements. IIIEE is interested in getting involved in environmental audits, but draws the line at Environmental Impact Assessments (EIAs) as 'too close to biological ecology'. As for the industrial side in the UK – including Scotland, which has sent visitors to the institute – he felt that quite abstract systems of management tended to obscure practical issues. As he put it, 'The Brits are much happier observing than intervening.' He referred to the (in his view) rather over-theoretical and now out-of-date work by Pearce and his colleagues in their 'Blueprint for an Environmental Economy', but, clearly feeling that fair is fair, he added that he respected John Elkington.

The international crops of students taught at the institute are its pride and joy. I was taken round the rather labyrinthine corridors and staircases of the carefully restored late 19th-century institute building and chatted to the students, usually two to three per study room. It was impossible to not to be impressed by the positive things that were said. The intake to the M.Sc. course in Environmental Management & Economics (conducted in English) is selective (about one in ten applicants get in) and each year several go on to the doctoral programme, all free of charge under existing Swedish legislation.

Interview with Arne Jernelöv,
(then head of the Swedish Advisory Board for
the Research Councils (Forskningsrådsnämnden))

The Advisory Board for the Research Councils (ABRC) had a key function and Arne Jernelöv was better informed about R&D at the national level than practically anybody else in Sweden.[27] 'Had' and 'was', because the system has changed since the interview (November 1999). Arne Jernelöv has moved on to the International Institute for Applied Systems Analysis in

Vienna. ABRC is now the Academy of Sciences, as established in March 2001.

But instead of consigning the whole interview to the paper recycling bin, it is worth tracing the background to the change. Some figures: Sweden spends more in proportion to GDP than all the industrialised countries; 4 per cent in 1997, which compares with about 3 per cent in the US and Japan. The 1996 figure for UK was 1.4 per cent. It is true that this includes R&D carried out by the disproportionately (for the size of Sweden) large number of very big firms, which contribute as much as two-thirds of the total: Ericsson, Telia, Volvo, AstraZeneca, PharmaciaUpjohn and so on. When it comes to state spending, Sweden lies close to the OECD average and, fairly typically, about one-third of the state cash is channelled into research council budgets. Most of the rest goes into direct support for universities. Also highly recognisable is the problem of co-ordinating the many (thirty-odd) public bodies with a stake in the R&D largesse. One example: how much to allocate to the crucial Joint Energy Authority, when it argued for extra research funding going into its re-orientation towards new, renewable sources of energy. Sweden is operating a system of 'eco-labelled' energy and there is a certain pressure on official bodies to be seen to be supplied from eco-sources. The authority was funded and the scheme seems to work, but not everybody approved of seeing money being spent on applied research carried out by a quango.

At the time of the interview, the government brief for the research councils was that while tertiary education's 'faculty-research' was free of topic guidelines, applied research should be adjusted to the need of social sectors. The advisory board's main function, apart from dealing with communication issues and applications for expensive items of machinery, was to co-ordinate interdisciplinary (in scientific terms) projects and projects that crossed borders of socially defined sectors, for instance economic inquiries into environmental projects.

However, there was a parallel funding stream, emerging from a restructuring of the welfare state's most laden coffers: the state pension funds. By the early 1990s these funds had become an embarrassment to the then ruling conservative coalition and were handed over to the Ministry of Education for the creation of research trusts, designed to be partly independent of the state. One of them was the Foundation for Strategic Environmental Research (MISTRA), with a budget that is about one-twentieth of the total spending on all university research. Membership of the MISTRA board meant real power, because not only strategic but also direct funding decisions were made at that level. A proportion of board members must have relevant research experience, but state and 'sectional' needs were also represented. Risk management, social science aspects and environmental law now form part of environmental groups of subject areas, but in some of these, MISTRA collaborates with the Humanities Research Council. In addition, after reversing some rather misguided tendering-out plans, the Swedish Environment Protection Agency is now spending serious money on research. The ABRC itself directed funds into environmental research, for instance the world-class project of the joint ABRC/EU space and environment research institute in the extreme north. It is mainly aimed at investigating levels of stratospheric ozone and climatic change.

We discussed some real, then current, cases of sectional research funding, including support of risk management studies. They were classic 'big enterprise' projects. One concerned the collapse of a motorway tunnel through a ridge made of unsteady Ice Age moraine.

The most burning question of the many raised by this undertaking to be answered: When do financial interests back down and allow best-practice research results to dictate progress? The biggest inter-Scandinavian project of all at the time of the interview was the Sweden–Denmark bridge over the Öresund Straits. Among the engineering studies and socio-economic forecasts, environmental impact assessments were also looked to for guidance. Evaluations of emission risks mattered. Should the bridge allow car traffic or only public transport? The conclusion was the old ferry-traffic on international waters escaped Scandinavian restrictions and relied on dirty, sulphurous diesel fuel. When catalyser-equipped, fuel-efficient Nordic cars took over from the shipping, emissions would actually fall. This clinched the decision, given the economic pressures in favour of car traffic.

Jernelöv backed the argument that environmental legislation in the 1970s and early '80s drove the sustainable development culture in Scandinavia. Everyone agrees that the social alarm was raised in Sweden and Finland by the environmental crisis caused by the waste from chlorine-based bleaching of paper pulp being pumped into the Baltic Sea. The paper mills fought back initially, but a united scientific frontline stand was established and, crucially, was backed by consumers. The entire cellulose-manufacturing base was changed. Things moved very rapidly after that, from eco-labelling at inter-Nordic as well as national level to new legislation on forestry management in which environmental and production goals are given equal weight.

The big issue then was also pushed into prime position by public concerns: a radical policy change for permitted use of chemicals that spread and/or accumulate in nature. I asked about the EU and its dislike of unilateral action by nation states, but Jernelöv answered with a smile that the new Environmental Commissioner was Swedish.

Further Comments

If all this sounds too good to be true – well it is, to some extent. Concurrently with the sustainable development successes there have been disappointments about many things, from profitability to excessive bureaucracy. The sectional research policy, with its emphasis on directive spending, has served environmental researchers well on the whole but has been under concerted attack, perhaps particularly from academic institutions. The main reason seems to be what higher-education-based researchers regard as financial neglect of 'free' non-applied projects in favour of a 'problem-solving' approach.

There is another, related argument, which says that trusts such as MISTRA and EU influence already have destabilised national control over research spending. The counter-argument is that a shift in the balance of power towards the academic research establishment will correspondingly damage the interdisciplinary and applied research base in Sweden, and take money out of defined channels into the faculty spending pots. The 'academic' redirection has taken place, and with it seems to have gone all mention of environmental research from the programme of the Academy of Sciences. The jury is out on what will happen next.

Meanwhile MISTRA is still going strong, and for those who like to believe in rational management of the world we live in, its current research programme is a model (see box):[28]

> Strategy for the risk management of chemicals; Land use strategies for reducing net greenhouse gas emissions; Abatement strategies for transboundary air pollution; Sustainable building – an ecocycle system in buildings and construction; Sustainable forestry in southern Sweden; The potential of pulp and paper production as an energy producing and truly ecocyclic process; Eutrophication cost-effective measures for the Baltic Sea; Pheromones and kairomones for control of pest insects; Microbial antagonism against fungi; Paths to sustainable development – behaviour, organisations, structures; Remote sensing for the environment; Swedish regional climate modelling programme; Mitigation of the environmental impact from mining waste; Soil remediation in a cold climate; Batteries and fuel cells for a better environment; Ångström solar centre; Sustainable coastal zone management; Swedish water management research programme; Food 21 – sustainable food production; Sustainable management of mountains; Sustainable urban water management; Soundscape support to health; Plasma enhanced reaction systems for environmental applications; Management of seminatural grasslands; Networks between MISTRA and scientists in developing countries, 1997–9; Network – young scientists, 1997–9

Research and Design – and Development

Are designers partly responsible for Scotland's failure to have a distinctive market profile in sustainable technology? I feel it is paying designers a compliment to include them, because it means that the contribution of design to the R&D process has been taken seriously. Too often people, including designers themselves, behave and speak as if the word 'fashion' was understood always to precede the word 'designer'. But designers tend to be the first to insist that their contribution is not to do with shapes and colours and oracular announcements of quick-change aesthetics, but about technical innovation that is aligned to creativity and sympathy with human needs. There is a brief discussion in Chapter 7 of what I saw as design letting the environment side down – especially the city environment – exemplified by the Design and Architecture year in Glasgow. Here I shall just list some ideas, each time posing the implicit question: Why not in Scotland?

From *Trespassers*, a collection from the Netherlands Design Institute:[29]

> RED-HOT TARMAC. Even when a road is busy and 50,000 cars use it during daytime, vehicle bodies cast their shadow on it for only four hours. The sun warms it. When the outer temperature on a mildly sunny day is 15 degrees Centigrade, the tarmac heats up to almost 50 degrees Centigrade . . . The idea to use tarmac to collect solar energy – its dark colour makes it suitable – is not new [but collecting it used to be thought to rely on expensive copper piping]. Now Ooms Avenhorn Holding bv, producer of polymer-modified bitumen, together with WTH vloerverwarming bv, underfloor heating specialist, have patented a new solution. It involves plastic tubing that can be put into the tarmac in one go, together with an underlay of an insulating layer of tarmac, usually there anyway . . . The system is even more promising than hoped. [Instead of just solving the icy-roads problems] 80 per cent of the

stored energy is going spare, and can be used for other space-heating, for example in buildings. One kilometre of road is sufficient to heat 100 houses.

From the same source, an attractive way of using wood, with enormous potential:[30]

16. *Wooden motorway overhead display by the Dutch Department of Public Works. The idea of using laminated wooden structures is now being extended to other signs and to crash-barriers.*
Reproduced by kind permission of Zwarts & Jansma

HIGHWAY WOOD. Motorway signs, crash barriers and portals for traffic lane signalling are usually made of steel [maybe because metal is associated with engineering progress and also a well-understood material]. The Dutch Department of Public Works is experimenting with wooden structures for traffic guidance applications, designed by the architectural firm Zwarts & Jansma. It appears to have many advantages [wood is renewable], needs little maintenance and has less embodied energy than metal . . . Wooden crash barriers are better, because no polluting zinc is involved.

Next, a more challenging quote:

TEXTILE MATERIAL FLEECE. Fleece is made in Patagonia from recycled plastic bottles – Between 1993 and 1996, Patagonia has diverted 54 million plastic bottles from landfills; in the US about 9 billion plastic bottles are produced annually. Each garment needs an average of twenty-five bottles. Simple calculation shows that just one day's worth of bottles amounts to an entire year's production of sweaters. The green argument is a bit shaky [until a new idea for using bottles comes up.]

Furniture design: Summary of specification submitted to a competition at the Design Museum, London by designer Åke Axelsson from Galleri Stolen, Stockholm. Personal communication:[31]

All models in the 'Spring' series [use only] four pieces of wood. The raw material can be utilised in a more efficient way . . . The textiles used are pure wool. Any leather comes from Swedish cows and is tanned with vegetable based products. The wood surfaces are finished with waterbased colours and treated with oil or wax. [No toxic products are used in the manufacture.] The simple and carefully thought out design of 'Spring' [furniture] aims to minimise the use of energy in production, transportation and distribution. The product can be shipped disassembled in order to reduce freight costs.

Sixty per cent of the [furniture] factory's energy needs are filled by using waste material from the manufacturing process. The remaining forty per cent comes from products acquired from a nearby sawmill. [The technique has been developed in collaboration with researchers at the Stockholm Technical University.] Chimney filters ensure that all Swedish standards for emission are met. Energy consumption has remained nearly constant during the last ten years, while production has quadrupled. The furniture factory is located close to a small lake with clean, portable water and with a healthy fish and crayfish population . . .

No water is used [any more] in the factory's operations and there are no effluents from the mill. The factory owns a rowing boat that is available for the employees' recreational purposes. Shipments from the factory go directly to the consumer by truck, which helps to reduce packaging to a minimum [and] shipments are coordinated . . . to reduce the number of miles driven . . . The factory is a member of REPA [the Swedish Recycling Association for Packaging].

When we purchased the factory ten years ago, the work force consisted of four employees. Today there are sixteen and the production has increased fourfold. This positive development is of the utmost importance for the population of the small town of Skirö . . . where the factory is located.

Notes

1. Organisation of Economic Co-operation and Development (OECD). Members are: Australia, Austria, Belgium, Canada, Czech Republic, Denmark, Finland, France, Germany, Greece, Hungary, Iceland, Ireland, Italy, Japan, Korea, Luxembourg, Mexico, Netherlands, New Zealand, Norway, Poland, Portugal, Slovak Republic, Spain, Sweden, Switzerland, Turkey, United States, United Kingdom. The OECD Convention requires member states 'to promote policies designed to achieve the highest sustainable economic growth . . . while maintaining financial stability . . . and contribute to the expansion of world trade on . . . a non-discriminatory basis' (Article 1). Also, 'to . . . liberalise world trade and

co-operate in bringing economic development of all the world's countries' (Article 2). www.oecd.org

2. 'On the Scottish Road to Sustainability?' by Kevin Dunion, in *Scotland: The Challenge of Devolution* by Alex Wright (ed.), 2000, pp. 188–203.
3. Ibid., p. 192.
4. Ibid., p. 193.
5. *The Green Capitalists. Industry's Search for Environmental Excellence* by John Elkington, 1987.
6. *Factor Four: Doubling Wealth, Halving Resource Use* by Amory B. Lovins, et al., 1994; and *Natural Capitalism. The Next Industrial Revolution* by Paul Hawken et al., 1999.
7. *Natural Capitalism. The Next Industrial Revolution* by Paul Hawken et al., 1999, pp. 176–7.
8. *The Green Capitalists. Industry's Search for Environmental Excellence* by John Elkington, 1987, p. 180.
9. *An Illustrated Guide to the Scottish Economy* by Jeremy Peat and Stephen Boyle, with Bill Jamieson (ed.), 1999.
10. Ibid., e.g. p. 4.
11. Ibid., pp. 133–5.
12. 'Using Economic Instruments to Improve Environmental Management' by N. Hanley and D. Macmillan, in *Scotland's Environment: The Future* by George Holmes and Roger Crofts (eds.), 2000, pp 92–103.
13. *An Illustrated Guide to the Scottish Economy* by Jeremy Peat and Stephen Boyle, with Bill Jamieson (ed.), 1999, pp. 80–3.
14. *Boyack Urges Scottish Businesses to Take the Sustainable Technology Route.* Scottish Executive Press Release, August 1999.
15. Sustainable Development debate introduced by Sam Galbraith. Scottish Parliament Official Report, 28 February 2001, Col. 19.
16. *Towards a Sustainable Future. Annual Review 1999.* Forward Scotland, 1999; and *Looking back on a Year of Looking Forward. Annual Review 2000.* Forward Scotland, 2000, p. 12.
17. Forward Scotland has seven people on the Board of Directors, all with other weighty commitments, and eight permanent staff. See *Looking Back on a Year of Looking Forward, Annual Review 2000*, inside front page.
18. Scottish Institute of Sustainable Technology (SIStech) advises on conference management from its website: www.sustainable.conference.co.uk. An inquiry produced a set of single sheets (sent 19 April 2001), each describing one project. These were: An Examination of Multi-Criteria Decision Making in Water Supply and Waste Water Systems; COWASTE (a Northern Periphery project, i.e. with unspecified partners from three Scandinavian countries) or The Development of Recycling and Utilisation of Construction Waste; Developing a Further and Higher Education Partnership in order to 'achieve strategic objectives . . . and positive engagement with the sustainable development agenda'.

19. Heriot-Watt University – Centre for Environmental Resource Management. www.hw.ac.uk/cerm
20. Central Research Unit, Research Programmes. In 1999–2000, CRU will commission just over £5.7 million of research on behalf of policy makers. The current research programmes are: Civil Law and Legal Aid 1997–2000; Criminal Justice Research Programme 1998–2001; Development Department 2001–02 (e.g. housing, urban regeneration, planning, local government, transport & road safety); Education; Environment Group Research Programme 2001–02; Social Work 1996–9; The Children (Scotland) Act 1995; Rural Policy Research Programme 1999–2000. www.scotland.gov.uk/cru/res-progs.asp
21. Agricultural and Biological Research Group (ABRG) in the Rural Affairs Department. www.scotland.gov.uk/abrg/research_programmes.asp and www.scotland.gov.uk/abrg/sponsored.asp 'Main sponsored bodies' include the Hannah, Macaulay Land Use, Moredun (animal infectious disease), Rowett and Scottish Crop Research Institutes, the Scottish Agricultural College and the Royal Botanic Gardens Edinburgh.
22. *Natural Capitalism. The Next Industrial Revolution* by Paul Hawken et al., 1999, e.g. 'Making the World', pp. 62–81, and 'Capital Gains', pp. 144–69.
23. EMAS information. http://europa.eu.int/comm/environment/emas/. International Standards Organisation information www.iso.ch/infoe/faq.htm
24. 'On Environmental Technology' ('Tema Miljöteknik') by Roger Thorén (ed.), in 'Miljö Rapporten', *Ekonomi +Teknik*, April 2000, pp. 13–17.
25. 'Status of Life Cycle Assessment (LCA) Activities in the Nordic Region' by Ole J. Hanssen. *International Journal of Life Cycle Assessment* 1999, 4 (6): pp. 315–20.
26. Folkecentret [for Renewable Energy] www.folkecenter.dk/ www.avalibrary.dk/Aarhus/folkecen
27. The framework for Sweden's state funding of research is similar to the UK one: Research Councils administer funds to Humanities and the science sectors (Medical, Natural and Engineering Sciences). The old National Board for Industrial and Technical Development (NUTEK) has been split into agencies for Business Development and Innovation Systems, and an Institute for Growth Studies. There are also various national trusts, including the Foundations for Strategic Research and for Environmental Strategy.
28. The Swedish Foundation for Strategic Environmental Research (MISTRA) was established in 1994 with a capital of SEK 2.5 billion, deriving from the former Employee Investment Funds. The income earned on the capital is used to support strategic environmental research. MISTRA's budget for activities in 2001 amounts to SEK 250 million. Article 1 of MISTRA's statutes: 'The Foundation shall promote the development of strong research environments of the highest international

class with importance for Sweden's future competitiveness. The research shall be of importance for finding solutions to important environmental problems and for a sustainable development of society. Opportunities for achieving industrial applications shall be taken advantage of.' www.mistra-research.se

29. *Trespassers: Inspirations for Eco-efficient Design* by Ed van Hinte and Conny Bakker, 1999, pp. 27–8.
30. Ibid., p. 28.
31. *Product Submission to Sustainable Design Award Competition 1999* by Åke Axelsson and Dag Klockby. Designer responsible Managing Director at Galleri Stolen. Submission for Design sense event, Design Museum, London. Personal communication.

9

Research and Development in the Countryside

Rural Development?

BY THE END OF the 20th century, 'Rural Development' had moved to the forefront of public consciousness in the UK, but the decisive push did not come from any obvious direction. The cause was neither agitation for land reform, nor the nature conservation debate, nor the many concerns about deterioration of natural resources, including bio-diversity. It was disease, on a huge scale, which stimulated an essentially urban public to sit up and pay attention.

London Review of Books is not an obvious source to use for information about farming, but it published an interesting and thorough study on the subject by its ex-staffer, Andrew O'Hagan, which he called 'The End of British Farming?'.[1] The '?' seems little more than a grace-note, because although Andrew O'Hagan can see some future for organic farm produce, he registers weary regret at the decline and further decline of 'old-fashioned' farming. The article summarises the results of talking to supermarket managers, MAFF[2] civil servants, several different types of farmers and owners of farm-related businesses, and reading up on relevant material, from Deloitte & Touche's *Farming Results* to historical documents. It is a tale about the near-total collapse of yet another core industry in Britain – after coal mining, steel-making, ship-building, rail transport and many others besides.

Perhaps many British farmers deserve the misfortunes that have increasingly beset the industry, even though others suffer unfairly. Andrew O'Hagan began one passage with: 'I set out on my own rural ride feeling sorry for the farmers,' but a little later noted the decades of ill-gotten gains and quoted a civil servant saying: 'It [feeding animal waste to cattle etc.] would never have happened on a traditional British farm. It is part of the newer, EU-driven, ultra-profiteering way of farming. And look at the results.'

Look, indeed. To Scotland, for instance, which has followed essentially the same pattern as elsewhere in the UK. In 1995, the aggregated farm income was £600 million, of which £400 million came in financial subsidies and other aid. Until now, Scotland's new rulers have had remarkably little to say on the subject of farming and rural development in general. Maybe the cumulative effects of crisis following crisis will start to have an effect after 2001.

In 1998, the devolved Executive clearly felt some action was needed. It launched a consultation exercise called 'Towards a Development Strategy for Rural Scotland'. Later that year, the results of the exercise (only about 280 respondents, that is just over half the number that got writing on the more arcane subject of National Parks) were summarised in terms so anodyne as to amaze even hardened readers of such documents. Just a short extract to give the flavour is surely enough: '[The consultation] told us that we were right to put people at the heart of sustainable rural development; . . . we must recognise the diversity of rural Scotland and that we will only make real progress on rural development if policy is developed and delivered as a coherent whole . . . This gives us the clarity we need to take forward our commitment to sustain vibrant local communities in rural and remote areas.'[3]

Some words however have crucial meanings beyond what is immediately apparent. For instance, 'diversity' must be understood as referring to the new Agricultural Holdings legislation, which includes proposals to 'permit wider diversification by farm tenants'. The discussion of the Land Reform Bill mentions other changes with real importance to people who live in the Scottish countryside.

Rural Development hardly figured at all in the now First Minister's (then Minister for Enterprise and Lifelong Learning) exercise in dynamic forecasting 'The Way Forward: Framework for the Economic Development in Scotland'.[4] Even the Council for British Industry (Scotland) called this document 'a little vague'. There have been some helpful announcements made since.[5] For instance, the Minister for Transport has recognised the problems of running a decent public transport system outside the larger cities and towns (a kind of reverse definition of 'rural') in an announcement of £18 million over three years. How little this is becomes clear when you look at regional allocations. Only three local authorities get additional funding in excess of £1 million – Aberdeenshire, Western Isles and Highland – and that is meant to cover not only better services, but 'infrastructure costs' and 'installing Liquefied Petroleum Gas facilities'. Perhaps the news that Scotland is consulting with 'Nordic countries' (Finland, Norway and Sweden) on rural development issues is more hopeful, though the outline is vague indeed: 'the purpose of the study is to collect and share information'.

So what is 'Rural Development' after the end of farming-as-we-know-it? It is obviously not an exclusively British concern, but shared by all of Europe. There are proposals for Common Agricultural Policy (CAP) reform in Europe's *Agenda 2000*, notably cuts in livestock subsidies and increases in support for small farms. They are relatively limited in scope and seem hard to understand and to operate fairly. The agenda notes that intensive methods cause environmental problems, but mildly refers to the need for 'environmentally sensitive' farming. I know of no one who has written more lucidly about the complex and fundamentally destructive effects of CAP on the environment than the journalist Graham Harvey in *The Killing of the Countryside*, who also argued convincingly that 'alone in Europe, Britain abandoned the very concept of a wholesome diet when it threw its peasantry off the land.'[6] After the first wave of CAP reform proposals in the early 1990s, MAFF comments indicated that they, too, understood the relationship between environmental decline on one hand, and production subsidies, intensive livestock rearing and use of agrichemicals on the other.

The insights are there, but progress is slow. Still, there is hope when the Scottish Minister for Agriculture can make statements like this, referring to the previously quoted 'frameworks for development':

> The *Agenda 2000* package of reforms of the CAP and the Structural Funds have played a major role – and I believe the Rural Development Regulation is a milestone. [It will lead to] a genuine rural development fund – in line with Executive thinking in this area . . . A new agri-environment scheme – the Rural Stewardship Scheme – will include [assistance for] afforestation of agricultural land and the marketing and processing of timber products [and for] farmers and their families to diversify new income-generating enterprises, or to expand existing diversified businesses. There are measures to assist with the marketing and processing of agricultural products . . . linked to proposals for capital assistance; and a scheme to support our less-favoured areas (LFAs) – which make up around 85 per cent of the Scottish agricultural land mass.[7]

I suspect though, that most of the 'measures to assist with the marketing etc.' are meant to emerge from the inscrutable European Union Rural Development Policy Programme called LEADER, reincarnated as LEADER II.[8] Carefully defining its terms as it goes along ('What do we call innovative action?'), the last document describing the details of this Europe-wide sprinkling of projects is not very reassuring. However, it makes interesting reading if only because it emphasises how the deterioration of rural livelihoods, not only related to farming, has affected all of Europe, from Greece to Finland.

One of the most socially precise and therefore most moving accounts of this change I have seen charts the decline of Jorwerd, a small Frisian community, in a book with the sub-title *The Death of the Village in Late 20th Century Europe*.[9] It is a remarkable book, for the humanity in the descriptions of the people in Jorwerd and the meticulous observation of the events, large and small, which led to the community's disintegration and death – or did Jorwerd really die? There is a sense of continuity that remains, after all the profound and deeply familiar changes that rip the countryside apart – the migrations from the village to the city and vice versa, the bungaloid estates and the ruthless second-homers, the here-today-gone-tomorrow industrial development. The countryside survives but only just. As the Preface says, Jorwerd is representative of tens of thousands of villages in Europe:

> [By 1995] in the Netherlands 3,000 farming businesses were abandoned each year. According to EU estimates, half the agricultural capacity in northern Europe was likely to have gone within one single generation . . . The environmental lobby should have been the natural ally of the farmers, but the opposite was usually true . . . So farming as a way of life fell quietly between two stools, more than once, caught between family and the marketplace, tradition and modernity. Outside in fields the lambs were bleating and the auctioneer's calls echoed: 'The cheese press? 15, 30, 40, 40, 40 . . . going once, twice . . .'[10]

In the UK, many of the constructive ideas about what to do next seem to originate in England. A well-informed and practical Englishwoman called Rachel Thomas has summarised her view of 'the way forward' in a paper called 'Refocusing Rural Development'.[11] She has none of the Scottish Executive's problem with explaining what she means by a 'rural area': 'a useful commonsense definition . . . the countryside is where rural landuses dominate [and which] includes settlements of farms, hamlets, villages and small towns of up to a population of 10,000'. She insists it is essential 'to identify the "rural concept" . . . because of the assumption by some people that in reality there is no such thing as a rural area and people's needs are the same everywhere . . . and also that the rural economy, and particularly farming, is no longer important'. These assumptions are of course not unheard of in Scotland.

Rachel Thomas has outlined the many practical problems of living in the countryside and diagnosed most of them, from lack of services to high production costs, directly or indirectly related to poor transport. In some areas of rural England, where 80 per cent of the land is farmland and only about 10 per cent woodland, there are profound effects of the deep structural changes faced by farming. If at least some 20 per cent of farmland had to go out of production, then something else must be done with it to prevent the countryside becoming all heritage industry. At various points, the paper mentions real, small-scale and local developments, from glass-making to fish-farming to IT services provision.

She makes the very important point that the concept of environmental capital has got to be 'converted into actions to sustain it' and that this activity can be turned into bases for community regeneration. It is the old dream of a 'greened' community culture made to sound rational. Forestry is one source of work that is coming back: 'There is a very real buzz around in the forestry world.' Rachel Thomas should know – she is the chairwoman of a woodland development in south-west England ('the forgotten quarter'), tucked in between Bodmin Moor, Dartmoor and Exmoor.

Trees are indeed seen by many as the new and environment-friendly cash crop. Other ways of providing enterprises in the countryside while maintaining the 'environmental capital' are renewable energy production, eco-technology across a wide front, including individual transport (e.g. fuel-cell driven) and organic fish-farming. The unstable oil prices, the decommissioning costs of nuclear stations and the fossil fuel-driven changes in the climate make both renewable energy plant and large-scale sustainable forestry seem more economically plausible. They are now benefiting from accelerating investment both by nation states and oil giants such British Petroleum and Shell. Continuous alarms about managing bio-diversity etc. should result in more mainstream jobs for scientists of all kinds and at all levels.

The alluring picture of a busy countryside is not new to Scotland, but even more than elsewhere, it has historical evidence of how hard it is to make the ideal into reality. The emptiness of the Scottish countryside is largely due to the massive, 18th-century exercise in rural development known as the Clearances, when overpopulation by black-faced sheep replaced overpopulation by impoverished farm workers in many areas of Scotland. At the same time, another pattern was initiated by the ducal family in Sutherland: grand investments in economic revival. This led to new textile manufacturing plans for countryside, a model village at Helmsdale with a newly designed harbour, a new fish-curing facility and several new branches of established firms from places like Leven and Leith. The results can be summed up in this abbreviated quote from T. C. Smout's detailed account: 'Unfortunately . . . the volume of employment hoped for never materialised . . . There was in the end practically nothing else for the dispossessed peasant to do except dig their potato rig.'[12] Then there was Lord Leverhulme's almost complete repeat of the experiment, with an equally depressing outcome, but this time on Harris and in the mid-20th century. And then there was opencast mining and investments in aluminium smelting and tulip plantations and many other grand schemes.

Maybe the main lesson to be learnt from past failures of this type is that development and renewal has got to involve the community from the start. This has indeed become the politically correct thing to say, although the motives might be mixed. When the community takes over, not one, but many people take the blame for failures and schemes are often cheap and limited in scope.

Community and Enterprise

Communities are, as we know, the lynchpins of regeneration plans, both urban and rural. The support schemes discussed in Chapter 3 included both the floating staff of people with community development training and a handful of organisations, notably Dùthchas, CADISPA and Forward Scotland. The general impression of the organisational base was of laudable ideas, weak resources and a tangled undergrowth of official verbiage. But maybe the Scottish Enterprise network is a more obvious place to look for hard information about how Scotland supports small-scale local developments.

The Scottish Enterprise web-site home page shows little with stated relevance to community development, but offers links to, for instance, the Land Fund and important innovations such as Information and Communication Technology (ICT) and e-commerce.[13] Although electronic communications are seen as top-rated necessities for isolated communities, the SE website does not say very much about them, except to provide a colourful link to the Community Network. This turned out to be a dysfunctional set of a further four links under the heading 'Connect Me to My Community' but taking you to Fife Direct, West Lothian Online, Ayrshire Communication Society (all local authority-sponsored) and STB (now known as Visitscotland, the Scotland-wide tourism quango). I tried the Ayrshire one, remembering the environmental content of the Ayrshire Strategic Plan: it had nothing about communities or the environment.

The reason become obvious after following the Land Fund-link to its administrative centre. It has been devolved to the super-LEC for the Highlands and Islands (HIE), and this is where the community action is. Now, this is a bit odd: 'communities' of course exist all across Scotland. Isolation from the centres of power is not a variable that is directly proportional to the distance from the Scottish Enterprise offices in Glasgow or the Scottish Executive in Edinburgh. To place the entire Land Fund, as seems to be the case – all £10 million of it – at the disposal of the HIE constituency would surely be unfair.

It is not only the Land Fund that seems to be corralled for the inhabitants for the Highlands and Islands of Scotland.[14] There is, for instance, what HIE calls its 'best known and most common tool [for] assisting community development activity', the Community Action Grant (CAG) scheme. CAGs assist (up to 40 per cent of non-public funding) 'development projects undertaken by properly constituted voluntary groups'. Another example is the project called 'Sustainable Development in the Highlands and Islands of Scotland', one of twelve initiated Europe-wide in 1997.[15] So far, the outcome has been – as far as I can ascertain – no actual action other than that of creating a mass of paper-work, culminating in a report which was remarkable only for its exceptional vapidity.[16] This 113-page document has been based on the advice of ECOTEC Research & Consulting Ltd (an agency of the huge Asia-Pacific Economic Cooperation or APEC). ECOTEC's 'guidance' centres round sixteen verbose points. After progressing through various initial stages, it indicates how a region can really get going or, at least, how programme managers can find something to do:[17]

> ECOTEC recommends that programme managers map the measures of the proposed programme onto the 16 areas of action in order to assess how the programme will contribute to sustainable development and to identify additional measures and projects to increase the programme's contribution.

The HIE site has better ICT links to the local communities than the SE site and the messages from individual communities are usually well designed and up-to-date. Still, one wonders what happened to TITAN, the 'pan-European consortium of Regional Service Providers and Industrial and National Public Network Operators' and its Scottish branch-line, HI-ways (Highland and Islands ICT-network). The idea was to improve the quality of life for citizens and Small and Medium sized Enterprises (SMEs) in rural areas. We know that it was meant to 'provide advanced telematics tools to enable efficient navigation to public information and interactive services, of use as business support, local and public service provision, lifelong learning, and community networking'. We also know that www.hi-ways.org is a result of the collaboration between the key quango HIE and 'various public sector organisations and agencies in the Highlands and Islands' including several LECs. It all sounded so good – so why is there no trace of it on the HIE website?

The 'vanishing ICT network' is a fairly common phenomenon, probably due to the same kind of enthusiasm that fired the dot-com business flare-up and some of the sillier forms of e-commerce. One can only hope for the survival of the University of the Highlands and Islands (UHI), perhaps the most ambitious community development project of all. It is an almost-virtual university created by linking a geographically and academically disparate set of institutions ranging from local FE Colleges to centres for applied research. Some of these institutions are funded by local authorities, others are privately backed (in some cases as incorporated bodies) and some are linked to powerful industries based on, for instance aqua-culture and, indirectly, oil extraction.[18] The Millennium Commission awarded the project well over £33 million in 1996 and the remaining £60 million of estimated start-up costs were collected from European and local sources.[19] The difficulty of balancing aims of academic credibility and social and geographical inclusiveness are thoughtfully discussed in the paper 'The University of the Highlands and Islands Project: a Model for Networked Learning?'[19]

UHI has taken the environmental and sustainability issues on board and is providing modules and courses in its Faculty of Environmental and Natural Systems Sciences. The emphasis is on using the region as a 'natural laboratory', and 'to develop well-rounded graduates with a mature understanding of the environmental pressures and scientific opportunities the region offers'. The range is almost improbably wide, from postgraduate courses in 'mariculture' to HN Certificates and Diplomas in game-keeping and equestrian management. Although not listed as an academic partner, Heriot-Watt's Orkney-based International Centre for Island Technology seems to be a collaborator in an interesting course on topics under the heading 'Sustainable Development and Environmental Management'.

Just a brief look at what is happening in Wales might serve as a reminder of what can be achieved. There are several important nuclei of change already in place, perhaps most importantly the Centre for Alternative Technology, established over twenty-five years ago and by now commercially sophisticated.[20] There are the National Botanical Gardens, organically run and innovatively refurbished and organised, and several areas supported by European finance.[21] Wales is beginning to move towards a viable ecological infrastructure, notably through the investment in renewable energy (it has 49 per cent of all windfarms in the UK). One model community development in the remote Dyfi valley was enthusiastically described in a recent BBC Radio 4 broadcast.[22] It had the inevitable partnerships, the multiple funding sources, the NGO supported research projects and the interest in ICT, but it had also established ventures with a commercial future, from renewable energy installations to farmgate food products to a new plant for turning locally produced wool into wall-cavity insulation.

The Dyfi project was good on renewable energy, not only in the form of windfarms and solar panels, but in the use of regenerating woodland as an alternative energy source. Continuous fuel-feed stoves can be supplied by shredding, chipping or pelleting sawmill waste and coppicing wood. The possibilities of community forestry and value-added wood products should be particularly attractive to many Scottish communities on their way to economic viability.

The Forest as Redemption and as Industrial Plant: Climate Change

Simon Schama writes at length about European and English forests in his wonderful book about the landscape in history, art, literature and philosophy. It is a heady mixture, but in the end I decided to lift just one brief quote, in which Schama explains about Joseph Beuys,[23] the German artist and guru:[24]

> He wanted, so he said, to practice 'Verwaldung': afforestation as redemption. 'It suggests making the world a big forest, making towns and environments forest-like.'

The world needs redeeming, and if trees won't do it, maybe nothing will? There is much that is practical about this inspirational thought.

In environmental management programmes, forests = CO_2 sinks, that is the main greenhouse gas, carbon dioxide, is taken up from the air and built into plant compounds during growth. Trees can be useful in other ways too, for instance by absorbing other kinds of urban air-borne pollution or by screening unsightly things. But it is the sink concept that appeals particularly now when an apparently definitive series of reports on pollution-induced greenhouse effects from the Intergovernmental Panel on Climate Change (IPPC) seems to have broken through the barrier of doubt. At last, government scientists and other delegates representing about 100 states have 'unanimously approved the *Summary for Policymakers* of the draft report [first of three sections due in 2001] and accepted it in full'.[25] The data are alarming enough to have persuaded even major oil-producing countries to accept the report, though the judgement on whether the forecasts are as dire as they look now must of course be left for historians in the future.

Meanwhile, IPPC has published a helpful volume on land use and afforestation though the prose is not exactly easy reading – an example is included in the Note.[26] It documents the value of CO_2 binding in forests and any other vegetation as long as it is not promptly burned or combusted by methane-emitting metabolic systems (e.g. people or cattle).

It is easier to grasp the complexities of the idea as set out by the ingenious *Datatown* designers, who created notional forests able to absorb the assorted emissions of people and machines in their city/country with 241 million inhabitants.[27] After excluding

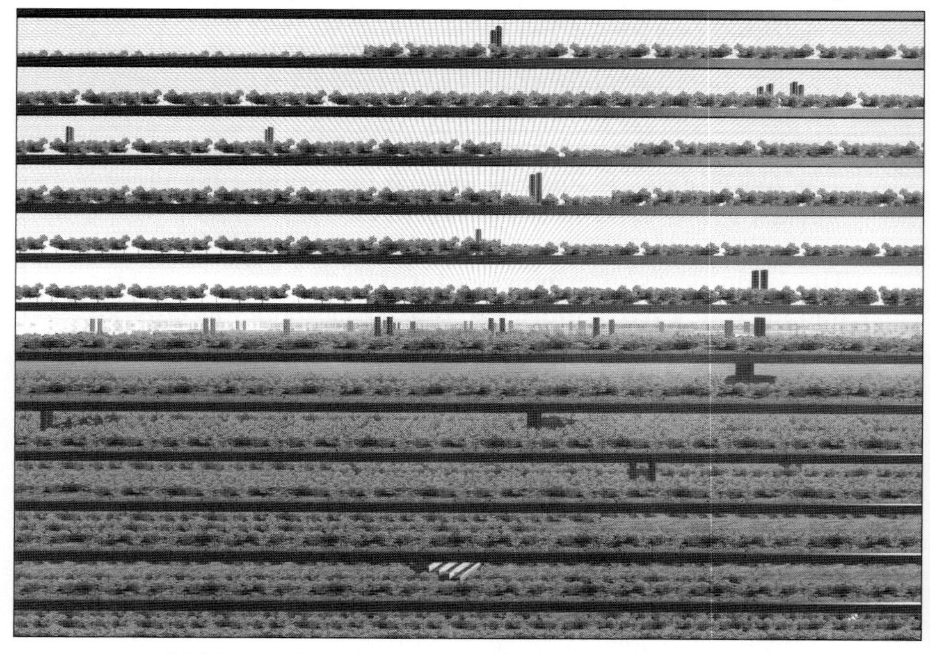

17. *Multi-storey Forest in Datatown. The carbon dioxide output of the industrial sector could be absorbed by a multi-storey forest (3,834 levels). The height of the 'forest block' would be over 100 km.* © MVRDV

agri- and horticultural land and parks, their 'CO_2-absorbing machine' – assuming that it was a reasonably attractive mixed forest – would need to occupy well over 4 million square kilometres, almost 140 times the size of the (present-day) Amazon region. The data-men solve the area problem by growing the CO_2-absorbing machine on 349 twenty-metre high floors in a vast forest skyscraper.

Although there is a sharply increasing interest in growing forests, there are also plenty of reports of forest destruction on a global scale. One recent example is the carving up of the Amazon rainforest in the name of Avanca Brasil or Advance Brazil.[28] Brazil is planning to spend $40 billion for several years from now to turn the Amazon into an industrial site; meanwhile the richest nations on Earth, the G-7 nations, have pledged $340 million in a non-recurring grant to support forest conservation in the Amazon.

It is hard even to imagine how much conserving the world's rainforests would require. It should be easier to see what can be done in the microcosm that is Scotland's forestry strategy. According to a recent DETR climate change document, Scotland's vegetation and soils constitute about 50 per cent of the UK carbon sink capacity and Scotland is planned to receive about half of the new planting envisaged for UK as a whole until 2010.[29a] Even the present planting scheme would apparently give credible 'carbon sequestration rates' of about 0.6 Mt C (million tonnes of carbon equivalent) per year and reference is made to the good tree growth rates in Scotland as justifying still further investment. In the context of global fuel policy, it is interesting to note that the oil giant BP Amoco has agreed to pay £10 million towards a 'sustainable forestry project' in Scotland with carbon uptake as a key objective.

The Scottish Executive's Environment Group has issued its corresponding document after a fairly modest consultation exercise.[29b] It notes that Scottish greenhouse gas emissions (a 'basket' of six gases, of which CO_2 is most relevant here) has remained unchanged throughout the 1990s, and worries away at the fact that unlike England, where emissions fell by approximately 10 per cent, Scotland's record does not look so good.

The UK government's stated commitment is to reduce CO_2 emissions by 20 per cent by 2010 (the UK's legally binding international target is a 12.5 per cent reduction by 2012).

Reafforestation of Scotland is not going to be the answer. While CO_2 fixation in assorted vegetation in Scotland is likely to be in the order of 0.4 Mt C in 2010, Scotland's 1995 emissions alone amounted to 24 Mt C per year. The problem is controlling sources of greenhouse gases – that is, energy consumption, including cars.

A Scottish Forestry Strategy: Caledonian Forest or Sitka spruce plantations?

In percentage of total land area, Scotland has more forest cover (16 per cent) than other parts of the UK, but it is still much less than the figure for Europe as a whole (33 per cent). The second half of the 20th century saw planting increase at a tremendous rate, driven first by post-war ambitions for timber self-sufficiency and later by a combination of an unstable rural economy and a intricate system of grants and tax incentives. The proportion of new forests to be managed as environmental and recreational assets also grew, but more slowly at first.

The state owns 38 per cent of Scottish forests and the rest (some 800,000 hectares) is in private ownership. The public forests are managed by Forest Enterprise Scotland, a commercially orientated and semi-independent off-shoot of the UK Forestry Commission Scotland FC(S), which functions as the Department of Forestry within the Scottish Executive. FC(S) is a policy-making, grant-channelling ministry controlling woodland management. FC(S) devolved its research function, too, and the result is called Forestry Research.[30]

FC(S) has issued a document called *The Forests for Scotland. The Scottish Forestry Strategy*, a strategic plan drafted in the conciliatory – even gentle – style of the modern civil servant, full of regard for partnership and proper consultation.[31] It has sections called 'The Challenge of Integration' and 'Realising the Vision'. 'The Strategic Directions' is a compendium of reasonable views on what forestry should be about (see box):

1. Maximising value
2. Creating a diverse forest resource
3. Making a positive contribution to the environment
4. Enjoying trees, woods and forests
5. Helping communities to benefit

Interestingly, the *Forest Strategy* focuses more on 'quality of life' aspects of forests, than on industrial or carbon-binding ones, with what I thought were overtones of a favourite Scottish idea, the Caledonian Forest. T. C. Smout dismisses the more extravagant versions of this.[32] However, it is a historical concept with deep roots. Like other partly voluntary schemes, it seems likely that the Lottery Fund-supported, imaginative and successful 'Millennium Forest for Scotland'-project is a re-enactment of the Caledonian Forest dream.

But although woodlands and forests – or a tree (or two) on its own – have value and significance on a spiritual plane, it is economic value that must form the core of the new forestry strategy. The timber and wood products market determines value, and people's enjoyment of trees can be converted into tourism and recreation. The *Forest Strategy* is bafflingly unspecific on the economic aspects of planting, maintaining, felling, transporting, manufacturing and selling timber products at any level of value-added refinement. Generally, there is a lot of uneasiness directly related to the sharp fall in timber prices in the competitive international market during the late '90s. In fact, the lack of quantification and tactics seemed the most noticeable things about the *Strategy*.

Had I missed something – was there something about this report that I had failed to grasp? What had the respondents to the two strategy consultation exercises (June and October 2000) said? The FC(S) was helpful and inviting when I asked to have a look. The consultation responses filled two towering piles of box-files, and I was also given the chance of a brief talk with the man in charge, Richard Broadhurst. He emphasised the extensive discussions carried out prior to the consultation exercises and the broad representation of stakeholders on the steering group. In response to a question about the lack of quantified processes and targets, he explained that the next step was to wait for data analysis, which was being carried out by an independent firm of consultants in Edinburgh.

Reading through as many written comments as possible from the later consultation, my first reaction was awe at the expertise and energy that had gone into the process, both by the respondents and the FC(S) staff. Most of the respondents were well-informed, articulate and prompt. Set out in the box below are the most consistent views amongst the comments I read.[33]

- Consultation process not working, since previous round of comments have not led to any changes in the revised document;
- Insufficient or no strategic targets, e.g. of land transfer into forestry, relative proportions of types of tree-cover etc.;
- No explicit financial strategy; no costings or time-scales, neither overall nor with regard to listed partners;
- No clarification of grants and other support e.g. for land transfer and agri-environmental schemes;
- Little or no indication of how to integrate policies for agri- and silviculture;
- No indications of framework for co-ordinating national, regional and local forestry strategies;
- No explicit statements on timber transportation logistics;
- No explanation of which 'added-value' process will be prioritised, to mitigate variability of bulk timber markets;
- Not enough emphasis or insight into private estate management;
- No reference to need for local foresters;
- No suggestions about how to sustain soil fertility and maintenance;
- No indication that traditional commercial forestry policy will be re-evaluated, e.g with regard to landscapes and planting densities;

- No clear policy of deer management;
- No plans for urban areas, community woodland, designed landscapes etc.;
- No clear statement of nature conservation goals in forestry context;

Such frustration is a not uncommon effect of consultation exercises. There were two types of comment that seemed to me particularly critical, in both senses of the word. One was: 'The consultation process is not working, since the prior consultation on the same subject did not lead to any changes in the [next] revised document'; and the other: 'There has been no explanation of which "added-value" processes will be prioritised . . .' The latter refers to crucial processes that turn timber into specialised – 'value-added' – structural and wood products.

One 'way forward' was set out in a document called *Roots for Growth*, published just before the *Forestry Strategy* (October 2000) by the joint efforts of Scottish Enterprise and Forest Industries Development Council.[34] Together, they manage what is known as the Scottish Forest Industries Cluster, 'clustering' being favoured by the Treasury as a way of strategically grouping resources in manufacturing for mutual support. The document was well received on the whole. Compared with the *Strategy*, it was better focused and quantified, sharper and more detailed in its criticism of weaknesses, including the need (though there was no discussion on how to meet it) for improving policy integration and both soft (research, training) and hard (transport) infrastructure provision. However, environmentalists were just about as unhappy with this as with the *Forest Strategy* document. Rob Edwards in the *Sunday Herald* summed up their response as 'business as usual' (quoting the views of Reforesting Scotland on the *Forest Strategy*) and 'The essentials of the deal were cooked up in a pre-emptive strategy for the forestry industry [*Roots for Growth*] agreed by [FC(S)] and Scottish Enterprise months previously.'[35]

In fact, developing woodland is something carried out across the board, from the huge Millennium Forest of Scotland – now concluded after five years and £11 million spent planting 8,000 hectares of mixed woodland – to small projects run by voluntary groups or individuals with a bit of land to spare. There is a bewildering range of support available, best seen in the Royal Forestry Society's remarkable booklet *Tr££s. Grants, Loans, Sponsorship and Advice on Trees and Woodlands in the UK.*[36] It lists thirty-three options under 'Government Forestry and Agricultural Grants and Other Government Grants for Trees' and thirty-six 'Regional Initiatives' as well as many other charity and industry-based sources of grants-in-aid and advice. The Woodland Grants system in particular has driven these smaller and medium-sized planting schemes, almost all with a heavy emphasis on 'amenity' – native, mixed, 'heritage' planting.

The environmentalists' enthusiasm for native woodlands and – in some cases at least – deep wild forests where even wolves and bears might roam freely, is in fact in stark contrast to the managed timber and wood products industry many want to see. So far, Scotland has been the prime region for UK forestry and to keep ahead there are large schemes in the pipeline, like the £400 million pulp mill plant currently being fought about by HIE and the Ayrshire and Grampian Enterprise Companies. But it certainly does not follow that smaller forest investors are any happier about the emphasis on redemptive tree-growing. A quote from the representative of the Scottish Crofters' Union to the North West Sutherland Council for Community Action for their third annual report in 2000 demonstrates the point: 'Many of us [in SCU] view with increasing uneasiness the spread of native woodland schemes and . . . question exactly what benefits the current policy is going to deliver. [It is] badly flawed . . . While they may be good for the environment and wildlife, they will never deliver the economic benefits that crofters are entitled to expect from forestry if their future is to be safeguarded.'[37]

Community Forests

The region of North West Sutherland, whose Council for Community Action's *Annual Report 99/00* is quoted above, was not picked at random. On one hand, Sutherland's Flow Country, the great peatlands (a proposed World Heritage site, currently with thirty-nine designated Sites of Special Scientific Interest), were subjected to the worst kind of ignorant commercial forestry by largely absentee owners in the 1980s. On the other hand, people living here look to income from forestry to ensure the survival of their communities, as remote and vulnerable to economic implosion and depopulation as any in the UK. For a few years now, the North Sutherland Community Forestry Trust (NSCFT) has been bidding for community ownership of Borgie forest.[38] It is an old forest, gifted to the nation by the Duke of Sutherland in 1917 for the benefit of returning servicemen, owned by the Forestry Commission and managed by Forest Enterprise. The NSCFT spokesman had great plans for a management agreement with Forest Enterprise, which would above all shift the timber cutting and haulage to local contractors, but provide also for local wood-product industries and even woodland tourism, mainly based on the sport shooting rights.

Plans like these are giving hope to many communities and are already being carried out in others. The management agreement idea was pioneered by the people in Laggan in the early 1990s and resulted in the Laggan Forest Trust. From early 1999, HIE has operated the Community Land Unit, now working alongside the Land Fund. It has supported 'a whole series of community [land] purchases', according to Dr Jim Hunter, the chairman of HIE.[39] He was writing in connection with yet another completed deal, this time in Abriachan on the shore of Loch Ness.

He saw forests and other upland as 'resources' for an expanding Highland economy: '[As people come] to set up businesses, there will be a growing demand for homes, for commercial premises, for places where people can relax and enjoy themselves.' This sounds more like economic development plain and simple, rather than environmentally as well as financially sustainable management. The environmental aspects of landownership changes are of course nowhere near resolved yet. The forest at Abriachan is not huge (540 hectares), but the largest community-owned forest in Britain. The purchase and its future is getting a sympathetic write-up by Alistair Scott in his helpful, nicely written compilation *A Pleasure in Scottish Trees.*[40] Scott describes how, over almost five years of negotiation and preparation, the villagers were helped to find the money (FC(S)/FE, the original owners, sold it for a modest £152,000) from eight major sources of funding and numerous minor ones. The community has, he feels, 'a great deal of good fun' as they develop their plans, which include all the familiar ideas about local wood products and tourism.

The results sounded promising, at least in the short term, but the process of getting there has been cumbersome and time-consuming. The Abriachan way is not acceptable as the route to restoration of Scotland's rural land (let alone its urban plots) to community management or viable small-holdings. The unsolved problems are piling up not least in the area of production management, as emphasised by Syd House in his interview later in this chapter.

There is an international background to community forests and it might be useful for Scotland to look carefully at the collected experience. The World Conservation Union (IUCN) has produced a wealth of evidence for global changes in forest management and the Overseas Development Institute has a whole series of Rural Development Forestry Papers.[41, 42] The ODI papers include evidence relevant to 'urban, industrialised' countries. Lars Carlsson emphasised this in his precise explanation of the share-holding system for 'common forests' in Sweden.[42a] It is a medieval form of share-cropping, but the assets are literally growing (only 70 per cent of annual biomass production is harvested) and the 'commoners' (the shareholders) have adjusted smartly to changing economic circumstances, for instance through mechanisation, digitised maps and production databases. The forest owners provide 'goods and services' – the 'services' include 'preservation of biodiversity' – but in spite of these high 'transaction costs', that is non-timber related costs, their investment in modern forestry is paying off.

The ODI papers also include a study of the background and possible future of three variants of community forestry in Scotland: the Laggan Trust, a jointly owned 'amenity wood' in the Borders and grant-supported crofting schemes in the north-west Highlands.[42b] The authors are based at the IUCN in Switzerland, where communally owned woodland has a long tradition behind it, and are very concerned about the nature conservation in Scottish forests. They gently but firmly warn against over-optimistic 'participatory schemes' but also against the deep suspicion about all these new bright ideas. It is a crofter with grazing-land as his priority, not an arch-conservative large estate-owner, who is quoted as saying 'If you want to plant trees, there's a traffic island down the road. That's the place for trees.'

Interview with Syd House

(Forestry Commission Conservator, Perth Conservancy. The Conservancy covers Central and East Scotland)

Syd House represents the modern, reformed Forestry Commission. It was reassuring to meet a senior FC(S) official with a genuine and wide range of interests in the landscape.

18. *A block of planted conifer comes into view as morning mist clears from the Cairngorm mountains near Glenmore.* © Colin McPherson

Scotland has got six FC Conservator posts. The Conservators have key roles in regional forest direction and management. Syd House describes their many-faceted function in terms of 'a regulatory and promotional brief on the lines of a technical Civil Service, rather than hands-on forest management', but emphasised that it was helpful to have – as he himself does – a wide range of operational experience. His territory includes Angus, Perth and Kinross, Stirling, Loch Lomond and the Trossachs and large stretches of the Central Belt, starting at the Firth of Forth and ending on the outskirts of Glasgow.

The post-war FC became synonymous with deep-ploughed hillsides either growing dull blocks of tightly packed conifers or scarred by clear-felling. The policy changed, gradually at first and then markedly during the 1990s. Syd House identified himself with the *Forest Strategy* and mentioned as an example his own work with a local NGO, Tayside Native Woods. It gets funding from various sources, including Europe, for their lead project of improving and extending mixed native woodland on Tayside. Like the growing number of similar schemes, lead by the Millennium Forest, the main idea is not to provide for future timber production. Instead there is a mixture of objectives, both structural (flooding and erosion prevention), economic (tourism) and biological (wildlife and plant habitats). Syd House does not agree with the critics of the *Forestry Strategy*, who accuse the FC(S) of 'carrying on as before' behind a shelter of conciliatory verbiage. He pointed to the commission's good environmental track record of using both its regulatory powers and its administration of the system of grants, notably the Woodland Grants Scheme to encourage sensitive afforestation. The FC's regulatory powers are wide-ranging and their consent is required for new forestry, deforestation and forest road building under the Environmental Impact Assessment legislation. It can also (but usually does not) intervene in larger (>0.25 ha) urban planting schemes.

The advisory role is not getting any easier, as large land holdings give way to greater community influence. I asked about a recent dilemma in Syd House's own territory, where the conflict, at least to the outsider, looked like a classic case of 'goodies versus baddies'. It concerned the village of Gartmore in a part of the Trossachs officially designated an Area of Great Landscape Value. The local people objected vigorously when it became known that a landowner proposed to plant 1.5 million Christmas trees on 500-odd acres of recently acquired farmland just inside the potential National Park boundary. Worse, the owner was a wealthy Danish corporate body, the Jensen Foundation, seen as secretive and influential and foreign. The Jensen Foundation already owns about 30,000 forestry acres of Scottish land. The objectors seemed to have impeccable environmental and democratic credentials. But the FC(S) had commissioned an Environmental Impact Assessment (EIA), which found no reason to suspect permanent environmental degradation. The rapidly turning-over plantations were technically regarded as a 'crop', not 'forest'. The EIA stated that the Christmas trees would not greatly change the appearance of the village's surroundings, nor would they prevent – as the villagers argued – wildlife from moving along a woodland corridor. The alleged transport implications, ruining recently installed cycling tracks and causing complete removal of an ancient bridge, were not correct either. FC did not come down on one side or the other. It recognises the legitimate interest of owners in managing their own land. Also, there had been a full consultation process, lasting one and a half years. Aspects of this case highlight the changing attitudes to land ownership and land use issues in Scotland's rural communities. As it happened, after a period when the Gartmore case was quite widely reported in the media and the temperature had been rising steadily, the issue was settled after a referral to the Forestry Commissioners: they turned down the Jensen Foundation's proposal. Generally, FC is keen to work with local interests, but there are areas in which differences of opinion were likely to occur.

One such area is innovation, both in forest management and wood products. FC has not only its own affiliated research establishments but also access to international research and development reports, whereas communities that chose not to become well informed might find themselves lagging. Another difficult set of questions related to income generation from forests is that communities might overestimate the productivity and profitability of growing trees. FC(S) is working closely with private enterprise and there is mutual understanding of the problems facing forestry.

Interview with James Dalhousie
(of Dalhousie Estates and Scottish Woodlands Ltd)
and Michael Osborne
(Technical Manager of Scottish Woodlands Ltd)

Scottish Woodlands Ltd is 'Scotland's longest established woodland management company'. It has got the important international standard (ISO) certificates for quality and environmental management, and offers its clients the option of managing their forests on the basis of the Forest Stewardship Council regulations. The Earl of Dalhousie is the Chairman of the Board of Directors, as well in charge of the Dalhousie Estates in Angus. He chairs the Natural Resources Committee of the Scottish Landowners Federation (SLF). Both men were open and generous not only with their time but also with core documentation about the industry.

A couple of questions about the Forest Strategy Document *Forests for Scotland* were briskly dealt with: the document was vague and lacked precise targets. The reason was not so much that FC(S) did not have the information, but that the Scottish Executive was still developing its forestry policy. Nationally, FC has needed bailing out by the Treasury because it has not been able to earn its keep by selling timber at a time when prices were falling on a competitive international market. It also had to provide timber to UK sawmills under existing contractual obligations. As it is, FC harvests its own forests to the limits of growth and Forest Enterprise has had to cut some of its 'amenity' programmes. Amenity forestry, that is forests managed for biodiversity or recreation, is not commercially viable at today's prices. The last report on FC by the Agriculture Committee of the House of Commons confirms the extent of the commission's financial problems: subsidies of £7.4 million from Westminster and £5.6 million from the Scottish Parliament in 1999–2000, and a estimated total of £18 million for 2000–01.[43]

Both men agreed about the lack of confidence in the future direction of forestry in Scotland, but felt that the devolved Parliament is showing a real wish to grapple with the problems. James Dalhousie added that in the absence of a definitive policy, conservation agencies had had too much influence. This was also due to past speculative forestry investments – notably the Flow Country case referred to earlier – that had caused bad feeling and a strong sentiment that Scottish land should be better used.

Michael Osborne discussed the role of FC in maintaining a steady level of planting and harvesting on a national scale. They own 35 per cent of all woodland – conifer and broadleaf – in Britain and over 40 per cent in Scotland, but because the bulk of the Forest Enterprise-managed forests are conifer plantations, FE produces 50 per cent of UK softwood timber. As the plantings from 1970s and '80s mature, output will reach a peak by 2020, but unless there is some form of value-added processing to enhance the sale prices, the industry will face serious trouble. I was given copies of the Forestry Industry's 1998 Handbook and the FC-produced *British Timber Statistics 1999* to take away.[44] There were strong doubts all round that the Scottish Executive would be ready to deal with a sufficient expansion of post-harvesting processing.

The quality of timber is another issue. Using intermittent felling rather than the clear-felling so disliked by conservationists; and partly or wholly natural regeneration by seeding from the remaining trees is called continuous cover forestry. The quality of the mature trees is high, but because the rotation time (from harvest to harvest) is in the order of 150–200 years, it requires large areas of standing trees. The UK overall lacks such areas, and this includes Scotland, in spite of being in principle a good place for trees in general and conifers in particular. Although huge plantings were made in the 1850s (the Seafield estates are quoted as having planted 31 million trees), the re-establishment of lost natural forest takes a long time.

I asked about another aspect of the 'added value': environmental certification. The Forest Stewardship Certificate requires a two-year record of sustainable forest management, and the onus of proof is on the applicant. It must include replanting, proof of care for wildlife and plant habitats, control of soil and waterway acidification and preferably, careful felling and transport practices. James Dalhousie said that while there were limits to the environment-friendliness of efficient timber production, the private sector was traditionally better at combining production with amenity forestry. Michael Osborne, who is a keen naturalist, referred to the many schemes already in existence, for instance Environmentally Sensitive Areas, Organic Aid (a modest government-run scheme to support conversion to organic growing), Countryside Premium and Farm Woodland Premium, that are all in one way or another aimed at expanding the 'agri-environmental' sector. He commented that delays seem inherent in the system and that the enthusiasm for planting is waning.

Will a National Park designation be a help or a hindrance for commercial forestry? Both James Dalhousie and Michael Osborne felt that the designating process was moving too slowly. The role of the FC in the context of National Park forestry was far from clear, although James Dalhousie said the Cairngorms Partnership had been a good forum for settling such policy issues, with its balanced representation of private interests, local authorities and state agencies. The Cairngorms Partnership was at the centre of the projects supported by the government's Forestry Challenge Funds in the area, the Forest of Spey and Deeside Forest, and worked closely with the similarly supported Grampian Forest.[45]

James Dalhousie felt that the landownership pattern in the area on the whole worked well. Large tracts of land are owned outright by organisations with conservation as their first priority. Criticisms of the large private estates, especially sporting ones, tended to be sweeping, prejudiced and short on pragmatism. The Dalhousie Estate was run as a balanced mixture of sporting moor and river, tenanted agricultural land, productive and amenity forest, conservation areas including an SSSI, and concerns such as a stately home and a garden centre. Of all these potential sources of income from land-related usage, only the sporting interest and the garden centre were consistently profitable. None of the three hill farms, all with over 2,000 acres of land, would have been economically sustainable without additional income, such as pensions or other employment.

Maintaining the high hills for non-sporting tourism alone would not make a contribution to rural economic survival comparable to the income from sport, which on his estate alone not only helped sustain the hill farmers but also five full-time gamekeepers and other staff. He ruefully described watching the slow and very expensive SNH-financed repairs of the paths up the eroded hillsides of Lochnagar. It provided jobs, as did for instance the also state-funded Ranger service, but was paid for by the tax-payer.

Would smaller scale or community ownership foster greater diversity of profitable land use? James Dalhousie referred to the Assynt purchase: the team was dedicated and effective, but troubled by lack of funds for land management. In the case of forests, it is the maintenance of plantations over a long time, and the direct relationship between the value of the land and the age of the trees that are the key considerations. A hectare of ten-year old conifers might be worth a third to a quarter of the price for the same area after another twenty-five years.

Forestry is a complex industry. It is also an industry that if well managed for long-term gain will be crucial to the development of Scotland's rural landscape. Can Scottish forestry be run both sustainably and efficiently enough to supply the needs of the native construction industry? A lot depends on enlightened investment in R&D. It is important to get the right wood for the right use and minimise milling waste and chemical treatments. Looking over the specifications for the Glencoe visitor centre, drawn up for the Gaia Group architects by the experienced and environmentally aware forester Bernard Planterose, the text is an implicit plea for better methodology, developed on an industrial scale.[46]

Notes

1. 'The End of British Farming' by Andrew O'Hagan. *London Review of Books*, 2001, 23(6): pp. 3–16.

2. UK Ministry for Agriculture, Fisheries and Food.
3. Towards a Development Strategy for Rural Scotland. Foreword by Donald Dewar. www.scotland.gov.uk/library/documents-w/ruralscot-01.htm
4. The Way Forward: Framework for the Economic Development in Scotland. www.scotland.gov.uk /library 3/economic/feds-00.asp
5. An £18 million funding package will provide major benefits to rural communities throughout Scotland. www.scotland.gov.uk/news/2001/03/se0734.asp. Scotland to Share Rural Development Experience with Nordic Countries. www.scotland.gov.uk/news/releas99_3/pr0663
6. *The Killing of the Countryside* by Graham Harvey, 1997, p. 121.
7. Scottish Executive Working with Europe to Address Challenges Faced by Rural Scots. http://www.scotland.gov.uk/news/2000/08/
8. LEADER [EU Rural Development] Innovative Actions. www.rural-europe.aeidl.be/ rural-en/action/
9. *Jorwerd. The Death of the Village in Late 20th Century Europe* by Geert Mak, 2000.
10. Ibid., Epilogue, pp. 242–67.
11. 'Refocusing Rural Development' by Rachel Thomas. Second lecture of two, given at the Royal Society for the Encouragement of Arts, Manufactures and Commerce. www.sage-rsa.org.uk/lectures/texts.refocus_rthomas2
12. *A History of the Scottish People 1560–1830* by T. C. Smout, 1998, p. 333.
13. My Community. www.scottish-enterprise.com/connect/my_community/
14. Land Fund (Highlands and Island Enterprise). www.hie.co.uk/welcome.asp.LocID-hiestrpricoacolslf Strengthening Communities. www.hie.co.uk/welcome.asp.LocID-hiestrint
15. The project Sustainable Development in the Highlands and Islands has been funded by DGXVI of the European Commission (EC) and has been undertaken by a partnership of local funding partners – all ten of the Highlands and Islands Enterprise (HIE) Local Enterprise Companies (LECs), Scottish Office Development Department (SODD) and Scottish Office Agriculture, Environment and Fisheries Department (SOAEFD). HIE acts as project manager and expert support is provided by Environmental Resources Management (ERM). The project is one of twelve pilot projects across Europe funded by DGXVI to assist with the integration of sustainable development into current and future structural funding programmes. The Highlands and Islands project is the only Objective 1 study, the other eleven being conducted in Objective 2 areas.
16. *Sustainable Development in the Highlands and Islands of Scotland.* Report from Highlands & Islands Enterprise, Reference 5241, 6 October 1999.
17. Programme managers are: North Atlantic Fisheries College, Shetland College, Orkney College, North Highland College, Lews Castle College, Inverness College, Sabhal Mòr Ostaig, Moray College, Highland Theological College, Seafish Aquaculture, Argyll College, Dunstaffnage Marine Laboratory (SAMS), Perth College.

18. Funding partners include: the European Regional Development Fund under the Highlands & Islands Objective 1 Partnership Programme, also North & West Grampian and the Rural Stirling/Upland, the Scottish Executive, Highlands & Islands Enterprise, Highland Council, Shetland Islands Council, Orkney Islands Council, Western Isles Council, Moray Council, Perth & Kinross Council, Argyll & Bute Council, Highland Health Board and the HIU academic partners.
19. The University of the Highlands and Islands Project: a Model for Networked Learning? by Veronica Adamson and Jane Plenderleith. http://collaborate.shef.ac.uk/nlpapers/adamson-p.htm See also www.uhi.ac.uk/publications/StratPlan00-04.pdf Strategic Plan
20. Centre for Alternative Technology. www.cat.org.uk
21. CYMAD is a company that encourages communities in Meirion, Arfon and Dwyfor. CYMAD's main focus is the administration of the eight Welsh areas assisted by the LEADER II programme (LEADER = Liaison Entre Actions Developpement de l'Economie Rurale), a European initiative to promote rural areas. CYMAD is a non-profit making company limited by guarantee.
22. Beyond the Broadcast. Changing Places. www.bbc.co.uk/education/beyond/factsheets/changing5_home.shtml
23. 'Art into Time: Conversations with Artists'. Joseph Beuys interviewed by Richard Demarco. *Studio International 195*, 1982 (996) p. 47.
24. *Landscape and Memory* by Simon Schama, 1995.
25. Climate Change 2001: The Scientific Basis by Intergovernmental Panel on Climate Change (IPCC) Working Party I, Cambridge University Press, 2001. www.ipcc.ch/
26. *Land-Use, Land-Use Change and Forestry* by R. T. Watson et al., 2000. The IPCC Definitional Scenario yields estimates of average annual accounted carbon stock changes from afforestation and reforestation in Annex I Parties from 2008 to 2012 of 7 to 46 Mt C yr^{-1}. This would be offset by annual changes in carbon stocks from deforestation of about -90 Mt C yr^1, producing a net stock change of -83 to -44 Mt C yr^1. If hypothetically, for example, afforestation and reforestation rates were to be increased in Annex I Parties by 20 per cent for the years 2000 to 2012, estimated annual changes in carbon stocks would increase (from 7 to 46 Mt Cyr^1) to 7 to 49 Mt C yr^1
27. *Metacity Datatown* by MVRDV, 1999; www.archined.nl/010
28. 'Death Sentence for the Amazon. Scientists say $40 bn Project is Set to Destroy 95% of Rainforest by 2020 by Steve Connor. *The Independent*, 19 January 2001.
29. (a) *Climate Change: UK Programme. Section IV, Actions by the Devolved Administrations.* Document published by the UK Department for the Environment, Transport and the Regions, HMSO November 2000; (b) *Climate Change: Scottish Programme.* Document published by the Environment Group, Rural Affairs Department, Scottish Executive,

November 2000. See www.scotland.gov.uk/climatechange

30. Forestry Research went independent in 1997. Its bases in Scotland are the Edinburgh area (Northern Research Station & Woodlands Surveys) and Dumfriesshire (Technical Development Branch). It also runs a network of ten field stations mainly used for monitoring and field trials.
31. *Forests for Scotland. The Scottish Forestry Strategy*. Forestry Commission, Scotland, November, 2000.
32. *Nature Contested. Environmental History in Scotland and Northern England since 1600* by T. C. Smout, 2000, pp. 37–9.
33. The Woodland Trust; R. D. G. Clarke, Edderton Farm; Ayrshire Council; Scottish Natural Heritage; Ramblers' Association Scotland; Lord Dalhousie, Dalhousie Estates; The Buccleuch Estates; Continuous Cover Forestry Group; Scottish Environmental Protection Agency; IM Forestry – Sustainable Forest Management; Gordon Timber Merchants; Timber Growers' Association; Macaulay Land Use Research Institute; Royal Scottish Forestry Society.
34. *Roots for Growth. A Strategic Framework for Action for the Scottish Forest Industries*. Scottish Forest Industries Cluster working document, Forest Industries Development Council & Scottish Enterprise, 2000.
35. 'Forestry's Millennium Doom: Woodland Strategy is a Damp Squib' by Rob Edwards. *Sunday Herald*, 31 December 2000.
36. *Tr££s. Grants, Loans, Sponsorship and Advice on Trees and Woodlands in the UK*. Booklet by the Royal Forestry Society. Also: www.rfs.org.uk
37. Third Annual Report by The North West Sutherland Council for Community Action. www.cali.co.uk, annualreport.pdf
38. 'Rooting for Future in the Forest. Hard-pressed Villagers to make Case for Ownership of Clan Lands' by David Ross. *The Herald*, 15 January 2001.
39. 'People are the Key to Land Reform. A Highland Community points to the Way Ahead' by Jim Hunter. *The Herald*, 15 July 1999.
40. *A Pleasure in Scottish Trees* by Alistair Scott. A private project document, made available courtesy of Michael Osborne of Scottish Woodlands Ltd. Project supported by the Millennium Fund and the Forestry Commission, 2001.
41. *A Long-term Strategy of the IUCN Working Group on Community Involvement in Forest Management* by M. Pfoffenberger, 1996.
42. a) 'The Swedish Common Forests: A Common Property Resource in an Urban, Industrialised Society' by Lars Carlsson; and b) 'Thinking Politically about Community Forestry and Biodiversity: Insider-driven Initiatives in Scotland' by Sally Jeanrenaud and Jean-Paul Jeanrenaud, 1996–7.
43. *The Work of the Forestry Commission*. Report, together with the proceedings of the committee, minutes of evidence and appendices. House of Commons Agriculture Committee, printed 7 February 2001.
44. *The Forestry Industry Handbook 1998*. The Forestry Industry Council of

Great Britain, 1998; *British Timber Statistics: Statistics on British Timber harvested and used by Primary Wood Processing Industries in Great Britain 1987–1999*. Economics and Statistics Unit of the Forestry Commission, December 2000.

45. Highland Forests. www.foresters.org/uk.htm
46. *Report on Timber Specification for the National Trusts Scotland Visitor Centre in Glencoe* by Bernard Planterose, 1999.

10

What Is Left of Scotland's Landscape?

The Vanishing Landscape

I KEEP COMING BACK to the simple fact that commerce and the landscape – including the cityscape – are not at opposite poles of the green value system. The sense of opposition is due to a failure of imagination. Commercial decisions need to be informed and tempered by evaluation of the environmental impact and guided by an understanding of a whole 'eco-system' of causes and effects related to each product. Unless this 'producer responsibility' process is enshrined in the law as well as in the consciousness of every citizen, land- and cityscapes will lose out to commerce every time. The reverse is also true: letting commerce act against the environment will have negative economic consequences every time, if not immediately so in the medium or long term. Surely there is enough insight around by now – enough understanding of nature's and industry's processes and sufficient fear for the future – not to let lack of care lead to more crises?

It would entail a quiet revolution, and the time to 'change the system' is running out. The Scottish landscape is vanishing at the same time as its sources of wealth are being drained. Examples of potential trouble are so plentiful that I am spoilt for choice – there is still enough material left to fill several more books.

What is happening to the sea? It is a truism to say that Scotland's coastal landscapes are very fine or that its waters are productive of fish and seafood, oil and gas and many other things. The west coast panorama of islands, hills, straits and sealochs is superbly beautiful, wonderful for sport and a rich nursery for marine life. So rigs have been put in place, pipelines lain and lorry-loads of toxic debris dumped. The north-east coast harbour towns have changed dramatically without the planning authorities taking more notice than they absolutely had to. Fisheries of all kinds have expanded unchecked and fish- and seafood stocks have fallen with entirely predictable relentlessness. As with many other sectors of the Scottish environment, Auslan Cramb has described these processes quietly but masterfully in *Fragile Land: Scotland's Environment.*[1]

The uncontrolled expansion of sealoch aquaculture is one of the saddest examples of loss both of a landscape and a commercial resource. The truly inexcusable business practices are beginning to get the attention they deserve, thanks to infuriated individuals, mostly drawn from the shellfish farming industry and environmental NGOs, led by Friends of the Earth and Scottish Environment LINK. I summarised the main problems and questions in an email sent – after an promising initial contact – to a Scottish parliamentarian, known for a wide-ranging and critical interest in marine issues.[2] There were also some questions about the fishery industry. My questions received no response at all. The box shows part of the email text. It included other details, which are listed in the Notes.[3]

In June 2000 the Scottish Parliament's Transport and Environment Committee was asked by environmental groups for an inquiry into sea cage farming covering these areas:

- Farmed salmon production has increased four-fold (1989: 28,553 tonnes; 1999: 120,000 tonnes)
- The number of staff employed is falling (1989: 1,418; 1998: 1,307)
- The number of companies involved in salmon production has almost halved (1989: 176; 1998: 95)
- Foreign companies now account for over 70 per cent of salmon production
- Escapes of farmed salmon are up to almost 400,000 fish in the first five months of 2000

- The livestock is riddled with disease
- Coastal water are severely contaminated
- Is anything definite known about the public inquiry?
- Industry press releases contradict many of the above points. Does Parliament have any independent information, for instance on staffing or worth to the UK?
- Who is responsible for policing the industry?
- What are the differences between organic and 'normal' farming practices?
- How much money has been paid to the industry under support schemes and in damages?
- Are other aquaculture projects – for instance shellfish, white fish farming – supported? By how much?

The questions imply that there are other ways of farming fish, which by setting much lower targets for growth and yield, and hence for profits, could make the whole industry positively

19. *A fish farm in Argyll.*
© Colin McPherson

benign. The essential changes would be simple to police: lower stocking densities, less intensive feeding with additive-free feed and a wider range of farmed species. The tremendous potential of the sealochs for 'good' aquaculture has been dissipated and an outstanding environment degraded. Worse still is the official blanket of silence that has been in place for almost two decades.

The oil exploration business is another corporate sacred cow. Large-scale environmental protest erupted around the proposal to dump the *Brent Spar*, a giant floating storage tank, in the North Atlantic west of Ireland and Scotland in spring 1995. The operation was due to start just before a meeting of the North Sea environment ministers on measures to eliminate the discharge of hazardous substances into the marine environment. Protests, notably by Greenpeace, led eventually to partial dismantling of the structure.[4] The whole affair was complicated when the green agitators were accused of using inaccurate research data, but dumping hardware in the sea has ever since been preceded by a great deal of anxiety and a voluminous EIA report. I was lent a report from 1996 entitled *An Assessment of the Potential Impacts of some Possible Options for the Decommissioning of the Maureen Alpha Production Platform, North Sea*.[5] It turned out to be a good example of what might be called the commissioned research-effect. Most of the report was thorough and careful, but the executive summary was oddly vague on detail and quick to come down in favour of deep sea disposal. The secret lay in the assessment methodology, meant to serve as the means of weighing factors for and against different procedures: it was an ineffective instrument. This was crisply described in the 'report on the report' by the husband and wife team Tasso and Margaret Elephtheriou.[6]

The outcome of 'the Maureen affair' is a stand-off, with – from an environmental point of view – reasonable implications. An expert, who shall remain anonymous but whose views have authority, commented: 'In reality, following the debacle that was the *Braer* disposal, it has been conceded by almost all of the original opponents of the deepwater disposal option, that this remains the Best Practicable Environmental Option (BPEO) until one takes into account the presentational aspects i.e. the public perception of the operation.' Some further comments in their personal communication are quoted in the Notes.[7]

Stories about environmental bad behaviour circulate quietly in the oil and gas industry, but even the most well-publicised cases are not handled according to the public interest. The *Braer* oil tanker break-up off the Shetland coast is a case in point, as re-emphasised by a recent investigation.[8] New, specific and potentially damaging information came to light thanks to 'a whistle-blower', who made publicly available a '2.5 inch-thick bundle of documents' including the 'previously unavailable logbooks from the *Braer* and from Statoil's Mongstad oil terminal in Norway'. What, one asks oneself helplessly, are public inquiries for? How can the public interest defence function, if it takes eight years and a mole in the system to produce basic evidence in an uncensored form?

Transport is another sector at war with the landscape. Again, Auslan Cramb gives it proper attention in *Fragile Land*.[9] As the landscape retreats further and further away from most Scottish citizens, the advancing army of cars is lead by the generals from the road construction industry and old enemy fortresses like Tarmac Ltd. This has happened all over the industrialised world, but there has been a fight-back in favour of the environment and of the industries that manufacture and run public transport systems. In Scotland the response has been weak. David Begg – Professor of Transport Studies in Aberdeen, government adviser and Chairman of its Commission on Integrated Transport – has said that the Scottish Executive is like 'a heroin addict' hooked on tarmac. Two unpublicised talks given by David Begg in Scotland were quoted in an article by Rob Edwards in the *Sunday Herald*.[10] The immediate subject matter was a new £500 million investment in motorways, including £250 million on an M74 extension into Glasgow. Characteristically, lack of proper evaluation was a key criticism. From Begg's present UK-wide perspective, it seemed that 'the climate for radical measures to steer away from car use to public transport is more hostile in Scotland. It is more conservative. There is a lack of commitment to the environment and a lack of commitment to public transport.'

The Skye Bridge story is another deeply worrying example of the distance between rhetoric and reality. Reported by George Monbiot, it is a case of corporate rule so blatant that it became one of the key examples in his book *Captive State: The Corporate Takeover of Britain.*[11] It is the long story of a still (the bridge project was completed in 1995) unresolved conflict between the infuriated, resourceful local population and the camp containing the Scottish Executive (initially the Scottish Office under Conservative control) and its business associates under a Private Funding Initiative (PFI) contract. Everything seems to have gone wrong. The cheapest construction option was chosen and the result is an aesthetic misfit and an insult to a beautiful place. The bridge is alleged to be too skimpy to cope with existing traffic and to have structural flaws. It is the only link to the mainland for the Skye people, because the ferry services were closed down, and they are charged excessive tolls (private car/van, £5.70 each way). The cry of 'Who benefits?' goes up – George Monbiot's answer is quoted below:[12]

> The bridge, according to the Miller group, which built it, cost £25 m. The Scottish Office initially contributed £13 m. The European Investment Bank added £13 m. Private investors paid in £7.5 m, the contractors £0.5 m and a commercial bank £6 m: a total of £40 m. The bridge was 60% overfinanced. The only explanation the islanders could suggest is that the investors would make a great deal of money. The £25 m project will extract, for its private backers and the Bank of America, £88 m from one of the poorest places in the British Isles.

Waste management draws even with the 'traffic industry' when it comes to complex, challenging civic and commercial issues. In spite of sporadic outbreaks of collective anxiety, usually by people living close to a landfill or waste burning operation, Scottish citizens seem uninterested. Businesses, forced by the landfill tax and other charges, are working at waste minimisation schemes, but the impetus is not sufficient. According to Audit Scotland (part of the Accounts Commission, Scotland) in 1999–2000 only 5 per cent of all waste was recycled, an all-European low (including England and Wales). This missed completely the government's 25 per cent target, set in 1991 for the year 2000. Less than 2 per cent of household waste was used for heat and power schemes and the cost for waste disposal by councils amounted to almost £100 million.[13] We know that it is perfectly possible to base both large and small enterprises on recycled materials, but the ever-repeated defence for inactivity is that 'the products can't be sold at sufficient profit'. It is a vicious circle, broken only occasionally by an imaginative design and investment policy – for instance Habitat store's current promotion of furniture from car tyres.

And so it goes. The Scottish record in biotechnology includes many remarkable successes, but the industry's potential for damage is as spectacular as its predictions of good things to come. Scottish Enterprise (SE) should not be allowed to provide uncritical support for the activities of the key institutes and centres (much foreign, notably US, funding has already been injected into their work) and independent research should be properly financed by the state. The report that SE is helping to infiltrate biotech-industry sponsored literature into Scottish schools is too close for comfort to the 'corporate takeover' paradigm outlined by writers such as George Monbiot, Naomi Klein and Noreena Hertz.[14]

As we have seen, the country sports and tourism industries, let alone agriculture, are not reassuring stewards of the vanishing landscape. The agricultural crises have provided hard lessons about how destructive uncontrolled exploitation of environmental resources can be. No one seemed willing to extend the conclusions to the relatively mightier tourist industry. Instead, much has been made of the 'idea' that planning restrictions should be lifted and large stretches of countryside turned into recreational facilities. This could become just a new way to exploit what is left of the landscape.

The Assessed Landscape

How do people react to landscapes? Many do not care that

much, some use landscape features for sport or business, but almost all make value judgements. Not a trivial observation, this, because 'landscape beauty' has become a serious quality that can be quantified. Do not imagine it is enough just to say 'Nice view'. The people who tell the rest of us about the worth and value of landscapes belong to the professional group called landscape assessors. Sometimes they are very useful, because their expertise lies in skills like ability to survey and chart a site and knowledge about soil decontamination and water tables. In fact, their skills are meant to be absurdly wide-ranging – land-use, soil science, nature conservation, landscape gardening and so on. Assessors are familiar with the planning regulations, know how to find grant funding and are good about aesthetic values as well. The President of the Landscape Institute has gone on record welcoming 'proposals that ensure a carefully planned environment in recognition of the changing use of the countryside . . . These plans must be firmly founded on proper landscape classification and assessment.'[15]

But carrying out 'proper assessment' is not a walk-over. When the Harris landscape was assessed as to the advisability of removing an entire mountain, Rodel, as crushed aggregate, controversy was not far away. The long-established conservationist view was that this was a fine place, a World Heritage Site and a National Scenic Area. The quarrying company engaged a landscape assessor who wrote a fifteen-page assessment disagreeing.[16] The landscape was classified into six types ('Rugged hills and glens', 'Rocky indented coast' and so on), its geological history was sketched and padded out with a literature search. The latter confirmed what we were already starting to feel: this landscape was no great shakes. Why, even Sir Walter Scott was said to have described it as 'high sterile hills. [I have] never seen anything more unpropitious'.[17] Vegetation and land use in the area were mentioned, but at this point a theme had begun to develop that was repeated throughout. The theme was 'visibility'. Who would see the landscape and from where? Using a 'visual assessment methodology' (walking around looking) soon led to a conclusion: 'Quarrying would not be visible from most of Rodel for the great majority of the working period.' Would it be visible from elsewhere, say, from Skye across the straits? It was felt that since the weather in the area was often 'unpropitious' too, most of the time the quarry would not be easy to spot. The assessor concluded that she had 'identified the scale and enclosed landform of the Rodel site are robust' and that it was in her view 'well able to absorb the change [caused by quarrying]'.

I have quoted at length, not so much because I want to go on to argue that the conclusion was utterly wrong or that the assessor's presumptions were improbably pedantic. Rather, the quotes are archetypal examples of how professionals provide 'science-speak' for the use of their political or commercial masters. To dismiss the Rodel report as a professional gloss to justify commercial intent is fundamentally to criticise its methodology. There were no clearly stated criteria for ordering and evaluating what was being scrutinised, and there was no clear scale or range of measurements and no relationship set out between what was being investigated and the aims of the investigation.

It matters for Scotland whether the aim was to enrich a foreign-owned quarrying company or to sustain its own citizens

20. *Peat cutters on Bernera, Lewis.*
© Colin McPherson

in a depressed rural environment. If the latter is the agreed aim, then it was worse than frivolous to keep the local economy in suspended animation for a decade while the QCs argued about how noticeable a lost mountain would be. The argument that the damage would not be very obvious and no one would see it anyway is of course also a breathtaking reminder of how little ground there is for hoping that anyone actually would care one way or the other. It is unlikely that there will be much support in present day Scotland for the notion that the landscape is worth having for itself – existing on its own terms, irrespective of whether seen or not.

Lost and Found Again?

This book is simply the record of how I have tried to pursue something that has been worrying me: the disappearance of Scottish land- and cityscapes. Any concerned citizen could have done the background work. I wish I could say that I had found a general undercurrent of interest, strong enough to match the flow of images and verbal portraits of Scotland. But although there are many attempts to do the right thing, they have often led to results that are oddly ineffectual. Everywhere 'real life', as measured and valued in terms of immediate gains, have tended to be seen as different in kind from 'nature'.

Although there have been prescriptive passages throughout, I have found it hard to summarise these into end-of-book conclusions. Perhaps my main concern is that the Scottish bureaucracy seems to be running out of control, in both senses of the phrase. In spite of many honourable exceptions, there are far too many desk-bound managers, who are either not specialist enough or not given enough leverage to be able to manipulate actual events. The room for decisive action by both politicians and their civil servants is declining in proportion to the proliferation of agencies and funding mechanisms. The civil servants, even the most well informed and thoughtful, are kept wandering in the labyrinthine structures of partnerships and consultation procedures. Meanwhile, the pressures from outside the system are building and corporate agents increasingly dominate the social stage. Corporations are not by definition bad news, not at all – but they are, when all is said and done, motivated by the need to maximise profits, now.

Without a united majority, democracy can become inactivated by sheer inclusiveness. My scattered examples from other countries have shown that focused public demand for quite radical environmental policies – right across the board, from nature conservation to product design – can be answered through the conventional power structures (no need to riot!). It is still a mystery to me why the Scottish people do not 'speak up'. Is it lack of insight? Most Scots still believe in a pure landscape 'away in the hills', but are increasingly cut off from the actuality of living and working in the countryside. Still, this does not explain the often indifferent or poor city environments. Scotland is not overpopulated (like The Netherlands or England) and it is not directly subject to the misbehaviour of neighbouring states (like the nations in central Europe). Could it be ignorance? Modern Scots are comparatively (with other First World nations) ill educated in mathematics and the sciences.[18] Maybe environmental systems and the processes used for analysis are seen as too complicated to bother about.

But these comments are unfair. Working on this book has given me the privilege of meeting people who do care, and are knowledgeable and effective. They tended to share a gloomy view of the present, but many felt that hopeful changes are in the air. Interest is growing in important areas such as nature conservation, sustainable forestry, organic farming and eco-friendly building and manufacturing. The increasing investment in renewable energy and the expectation of new, more accountable forms of land-ownership and land-use are even more vital factors. Scottish educational institutions seem more interested in environmentally aware teaching and research. The environmental lobby is already well organised and fast moving, but now includes a politically represented Green Party.

The last of Colin McPherson's pictures is taken on the Mar Estate, until recently in the hands of foreign owners apparently intent on proving that non-natives are even worse for the landscape than the natives. It is now being meticulously restored by the National Trust Scotland. The surreal-looking image is a very real demonstration of how being cruel can be kind: the strange shapes are jawbones of deer, which are systematically culled to a level that will allow the landscape to recover and grow varied and wild in its own right.

21. *Jawbones of culled deer on a line – used by stalkers to count the number of animals which have been legally culled – at Mar Lodge on the Mar Estate, Deeside.* © Colin McPherson

Notes

1. *Fragile Land. Scotland's Environment* by Auslan Cramb, 1998, pp. 97–121.
2. The MSP Jamie Stone, the Liberal Democrat Party's fishery spokesman. He made a strong contribution to the debate on Sustainable Development (Scottish Parliament, February 2001). Later he resigned as fishery spokesman, after a crisis due to North Sea fish stocks being so low that new restrictions on fishing became necessary, either by decommissioning trawlers (Lib/Lab coalition) or temporary tie-ups (SNP, Con and Lib/Lab rebels).
3. The responses of the industry can be seen for example on its website www.intrafish.com; comments in a BBC Documentary (5 January 2001)

were unrepentant. A couple of later pro-industry press reports stated: 'Fish farming, an industry worth £300 million to the Scottish economy, provides jobs for 6,500 people' (Special report by Stuart Millar Scotland editor, *The Observer*, 7 January 2001) and 'The . . . industry employs about 6,500 people, with 70% living and working in remote areas . . . seen as vital to the economy of the Highlands & Islands.' ('BBC Attacked over Salmon Report' by David Ross, *The Herald*, 9 January 2001). The following hazards have been reported, in addition to concern about foreign companies, overstocking and escapes: toxic algal blooms (bad for shellfish etc.) in 1999 and 2000; various forms of shellfish poisoning linked to severe marine pollution from caged fish including 'fatty vomit' due to overfeeding; contaminated fish food, for example by titanium dioxide, DDT, dioxins; toxicity of fish management chemicals, for example invermectin, tri-butyl tin (against sea-lice infestations); outbreaks of Infectious Salmon Anaemia & Swimbladder Sarcoma Virus; caged fish suffocating in large numbers due to over-stocking.

4. www.greenpeace.org/globalnews/
5. *An Assessment of the Potential Impacts of some Possible Options for the Decommissioning of the Maureen Alpha Production Platform, North Sea* by AURIS Ltd (Aberdeen), 1996.
6. *The Maureen Platform Abandonment Study. Critical Comments on the AURIS Report* by Tasso and Margaret Elephtheriou, 1996. The Elephtherious's comments are detailed and the quotes are indicative: 'the Severity Ranking [of categories for effects of different procedures] consists of a numerical amalgamation of all identified factors [which] gives a spurious element of objectivity'; '[In] the Severity Analysis each listing of each identifiable effect, whether positive or negative, represents only one item of the total. Even in arithmetical terms this process leads to a manifest distortion.'
7. 'Deep water disposal has, to all intents and purposes, gone away, but it is still one of the options that must be considered during the decommissioning phase and must be discussed in the Decommissioning Plan . . . The BPEO loses quite a lot of its attractiveness – the thought of Greenpeace crawling all over . . . sends the accountants scurrying for their calculators to divine how much the sales have dropped that week! . . . This refers to the deep ocean (outwith the precincts of the North Sea) in depths in excess of 2000m. Conventional legislation, e.g. the Food and Environment Protection Act, precludes disposal in the shallow waters of the UK continental Shelf. The UK government cannot say it will never undertake a sea disposal operation in case it precludes a force majeure disposal operations such as was required for the Piper Alpha. '
8. 'The Truth about the *Braer*' by Jonathan Wills. *Sunday Herald*, 15 April 2001.
9. *Fragile Land. Scotland's Environment* by Auslan Cramb, 1998, pp. 174–93.

10. 'Executive Hooked on Tarmac, Says Roads Tsar' by Rob Edwards. *Sunday Herald*, 8 April 2001.
11. *Captive State. The Corporate Takeover of Britain* by George Monbiot, 2000, pp. 18–58.
12. Ibid., pp. 50–3.
13. Audit Scotland is part of the Accounts Commission, Scotland. The information is drawn from Performance Indicators Environmental services – Performance Indicators 1999–2000. www.audit-scotland.gov.uk/
14. *Captive State. The Corporate Takeover of Britain* by George Monbiot, 2000; *No Logo* by Naomi Klein, 2000; *The Silent Takeover: Global Capitalism and the Death of Democracy* by Noreena Hertz, 2001.
15. Landscape Assessment Essential for Countryside Revolution. Cabinet Office Performance & Innovation Unit Report on Rural Economies. www.l-i.org.uk/newsdec99
16. *Rodel Isle of Harris Proposed Extraction, Processing and Transport by Sea of Anthrosite* by Moira Hankinsson on behalf of Redland Aggregates Ltd, 1992.
17. Ibid., p. 6.
18. *An Illustrated Guide to the Scottish Economy* by Jeremy Peat and Stephen Boyle, with Bill Jamieson (ed.), 1999, pp. 110–13.

Bibliography

(Named authors and editors only)

Arnold, Matthew *Culture and Anarchy*. Macmillan & Co, 1882.

Athanasiou, Tom *Slow Reckoning – The Ecology of a Divided Planet*. Secker and Warburg, 1997.

Atkinson, Kate *Emotionally Weird. A Comic Novel*. Transworld, 2000.

Baird, Vanessa 'Green Cities' in *Green Cities. Survival Guide for an Urban Future. The New Internationalist* 313, June 1999.

Banks, Iain *The Bridge*. Macmillan Ltd, 1986.

Bannan, Mark, Adams, Colin E. and Pirie, David 'Hydrocarbon Emissions from Boat Engines: Evidence of Recreational Boating Impact on Loch Lomond'. *Scottish Geographical Journal* 116(3): 245–56, 2000.

Barnes, Julian *England, England*. Jonathan Cape, 1998.

Barthes, Roland 'Myth Today' in *Mythologies*, transl. A. Lavers. Vintage, 2000.

Boddy, Martin, Lambert, Christine and Snape, Dawn *City for the 21st Century? Globalisation, Planning and Urban Change in Contemporary Britain*. The Policy Press, 1997.

Boyd, William *The New Confessions*. Hamish Hamilton, 1987.

Broun, Dauvit, Finlay, R. J. and Lynch, Michael (eds.) *Image and Identity. The Making and Re-making of Scotland through the Ages*. John Donald Publishers, 1998.

Callander, Robin *How Scotland is Owned*. Canongate, 1998.

Castells, Manuel 'The Power of Identity', in *The Information Age: Economy Society and Culture*, vol. 2. Blackwell, 1997.

Carlsson, Lars 'The Swedish Common Forests: A Common Property Resource in an Urban, Industrialised Society', in *Rural Development Forestry Papers*. Overseas Development Institute, 1996–7.

Carson, Rachel *Silent Spring*. Crest, 1964.

Craig, Cairns *The Modern Scottish Novel: Narrative and the National Imagination*. Edinburgh University Press, 1999.

Craig, David *Native Stones. A Book about Climbing*. Secker and Warburg, 1987.

Cramb, Auslan *Who Owns Scotland Now? The Use and Abuse of Private Land*. Mainstream Publishing, 1996.

Cramb, Auslan *Fragile Land. Scotland's Environment.* Polygon, 1998.

Crawford, Robert *Devolving English Literature.* Clarendon Press, 1992.

Cronon, William (ed.) *Uncommon Ground. Rethinking the Human Place in Nature.* W. W. Norton & Co, 1996.

Curry-Lindahl, Kai 'The Cairngorms Natural Nature Reserve (NNR): The Foremost British Conservation Area of International Significance', in *Caring for the High Mountains: Conservation of the Cairngorms* by J. W. H. Conroy, Adam Watson and A. R. Gunson (eds.). Centre for Scottish Studies, Natural Environment Research Council, 1990.

Darling, Frank Fraser 'Wilderness and Plenty'. Reith Lectures 1969 BBC, 1970.

Demarco, Richard 'Art into Time: Conversations with Artists'. Interview with Joseph Beuys. *Studio International 195* 996: 47, 1982.

Dewar, Donald 'Scotland Ready to Meet Challenges of the Millennium', in *Corporate Scotland 1998/99*, Bill Magee (ed.). Johnstone Media, 1998.

Duncan, James and Ley, David *Place/Culture/Representation.* Routledge, 1993.

Dunion, Kevin 'On the Scottish Road to Sustainability', in *Scotland: The Challenge of Devolution* by Alex Wright (ed.). Aldershot, 2000.

Dunn, Douglas *Scotland – An Anthology.* HarperCollins, 1991.

Dunn, Douglas 'Divergent Scottishness: William Boyd, Allan Massie and Ronald Frame', in *The Scottish Novel since the Seventies: New Visions, Old Dreams* by Gavin Wallace and Randall Stevenson (eds.). Edinburgh University Press, 1993.

Dunnett, Dorothy *King Hereafter.* Michael Joseph, 1982.

Duthie, Niall *Lobster Moth.* Fourth Estate, 1999.

Ehrlich, Paul *The Population Bomb.* Ballantine, 1968.

Ekman, Kerstin *The Forest of Hours.* Chatto & Windus, 1998.

Elkington, John *The Green Capitalists. Industry's Search for Environmental Excellence.* Victor Gollancz, 1987.

Ellen, Roy and Fukui, Katsuyoshi (eds.) *Redefining Nature. Ecology, Culture and Domestication.* Berg, 1996.

Fallow, Jeff *Scotland for Beginners.* Writers and Readers, 1999.

Foley, Gerald *The Energy Question.* Penguin Books, 1976.

Fraser, Maurice (ed.) *Essential Scotland. 70 Perspectives on the New Scotland from Leaders in Their Fields.* Agenda Publishing, 1998.

Fyfe, Nicholas R. (ed.) *Images of the Street: Planning, Identity and Control in Public Space.* Routledge, 1998.

Gervais, Marie-Claude *Social Representations of Nature: The Case of the Braer Oil Spill in Shetland.* Doctoral thesis at London School of Economics, 1997.

Giddens, Anthony *Runaway World. How Globalisation is Reshaping our Lives.* Profile Books, 1999.

Girardet, Herbert *The Gaia Atlas of Cities. New Directions for Sustainable Urban Living*. Gaia Books, 1996.

Glendinning, Miles (ed.) *Rebuilding Scotland: The Postwar Vision 1945–1975*. Tuckwell Press, 1997.

Glendinning, Miles and Page, David *Clone City. Crisis and Renewal in Contemporary Scottish Architecture*. Polygon, 1999.

Gray, Alasdair *Lanark*. Panther Books, 1982. First published by Canongate Publishing Ltd, 1981.

Gray, John *False Dawn. The Delusions of Global Capitalism*. Granta Books, 1998.

Greig, Andrew *Summit Fever*. Hutchinson, 1985.

Greig, Andrew *Kingdoms of Experience*. Hutchinson, 1986.

Greig, Andrew *When They Lay Bare*. Faber and Faber, 1999.

Grieve, Robert *Grieve on Geddes*. The Sir Patrick Geddes Memorial Trust, 1980.

Halliday, Sandy *Green Guide to the Architect's Job Book*. RIBA Publications, 2000.

Hanley, N. and Macmillan, D. 'Using Economic Instruments to Improve Environmental Management', in *Scotland's Environment: The Future* by George Holmes and Roger Crofts (eds.). Tuckwell Press, 2000.

Hanneberg, Peter and Zettersten, Gunnar 'The National Park the Locals Stopped'. *Enviro* 17: 16–18, 1994.

Hanssen, Ole J. 'Status of Life Cycle Assessment (LCA) Activities in the Nordic Region'. *International Journal of Life Cycle Assessment* 4 (6): 315–320, 1999.

Harvie, Christopher 'Industry, Identity and Chaos', in *Scottish Affairs* 32: 1–14, 2000.

Harvey, Graham *The Killing of the Countryside*. Jonathan Cape, 1997.

Hawken, Paul, Lovins, Amory B. and Lovins, L. Hunter *Natural Capitalism. The Next Industrial Revolution*. Earthscan, 1999.

Hertz, Noreena *The Silent Takeover: Global Capitalism and the Death of Democracy*. Heinemann, 2001.

Hinte, Ed van and Bakker, Conny *Trespassers: Inspirations for Eco-efficient Design*. 010 Publishers, 1999.

Hogg, James *The Brownie of Bodsbeck*, Douglas S. Mack (ed.). Scottish Academic Press, 1976.

Holahan, Charles J. *Environment and Behaviour. A Dynamic Perspective*. Plenum Press, 1978.

Holmes, George and Crofts, Roger (eds.) *Scotland's Environment: the Future*. Tuckwell Press, 2000.

Hoyt, Eric *The World-Wide Value and Extent of Whale Watching*. Whale and Dolphin Conservation Society, 1995.

Hutton, Will 'Urban Regeneration, Architecture and the Stakeholder Society', in *City* 7, May 1997.

Ingold, Tim 'The Temporality of the Landscape', in *The Perception of the Environment: Essays in Livelihood, Dwelling and Skill.* Routledge, 2001.

Ingold, Tim 'Ancestry, Generation, Substance, Memory, Land', ibid.

Ingold, Tim 'Hunting, Gathering as Ways of Perceiving the Environment', ibid.

Jack, Ian 'The Repackaging of Glasgow', in *Before the Oil Ran Out. Britain in the Brutal Years.* Vintage, 1997.

Jacobs, Jane *The Death and Life of Great American Cities: The Future of Town Planning.* Penguin, 1961.

Jeanrenaud, Sally and Jeanrenaud, Jean-Paul 'Thinking Politically about Community Forestry and Biodiversity: Insider-driven Initiatives in Scotland', in *Rural Development Forestry Papers.* Overseas Development Institute, 1996–7.

Jenkins, Robin *Poor Angus.* Canongate, 2000.

Kelman, James *A Disaffection.* Secker & Warburg, 1989.

Kelman, James *How Late It Was, How Late.* Secker & Warburg, 1994.

Kennedy, A. L. *Everything You Need.* Jonathan Cape, 1999.

Kinnersley, D. *Troubled Waters. Rivers, Politics and Pollution.* Shipman, 1988.

Klein, Naomi *No Logo. (No Space. No Choice. No Jobs).* Flamingo, 2000.

Kuppner, Frank *A Very Quiet Street: A Novel of Sorts.* Polygon, 1989.

Lambert, Robert A. 'In Search of Wilderness, Nature and Sport: The Visitor to Rothiemurchus 1780–2000', in *Rothiemurchus. Nature and People on a Highland Estate 1500–2000* by T. C. Smout and R. A. Lambert (eds.). Scottish Cultural Press, 1999.

Laslett, Peter *The World we have Lost,* Methuen & Co. Ltd, 1965.

Laslett, Peter *The World we have Lost – Further Explored.* Methuen & Co. Ltd, 1983.

Legge, Gordon *The Shoe.* Polygon, 1989.

Liddell, Howard 'The McLaren Building: Design Objectives', in *Dynamic Insulation – Past Present and Future.* Gaia Architects Conference, 2000.

Liddell, Howard, Roalkvam, Dag and Nettli, Harald *Pore-ventilation: Sports Halls. A Study for the Scottish Sports Council.* Gaia Architects Edinburgh/Oslo (Norway) & the Scottish Sports Council, 1995.

Liddell, Howard *Sustainable Community Development Model.* Gaia Group, 1999.

Lovins, Amory B., Lovins, L. Hunter and von Weizsäcker, Ernst *Factor Four: Doubling Wealth, Halving Resource Use.* Earthscan, 1994.

Mabey, Richard *The Common Ground.* Arrow Books, 1980.

Magnus Magnusson Foreword, in *Rothiemurchus. Nature and People on a Highland Estate 1500–2000* by T. C. Smout and R. A. Lambert (eds.). Scottish Cultural Press, 1999.

Mak, Geert *Jorwerd. The Death of the Village in Late 20th Century Europe*. Harvill Press, 2000.

Marr, Andrew *The Battle for Scotland*. Penguin Books, 1992.

Massie, Alan 'The Scottish Genius', in *Essential Scotland*, 1998.

Matthew, E. M. 'The Role of the Nature Conservancy Council in Protecting the Cairngorms', in *Caring for the High Mountains: Conservation of the Cairngorms* by J. W. H. Conroy, Adam Watson and A. R. Gunson (eds.). Centre for Scottish Studies, Natural Environment Research Council, 1990.

MacAskill, John *We Have Won the Land*. Acair, 1999.

McArthur, Alexander and Long, H. K. *No Mean City. A Story of the Glasgow Slums*. Longmans, Green & Co, 1935.

McCrone, David, Morris, Angela and Kiely, Richard *Scotland – the Brand. The Making of Scottish Heritage*. Edinburgh University Press, 1995.

MacDiarmid, Hugh *The Complete Poems*. Brian & O'Keffee, 1978.

McEldowney, John and McEldowney, Sharon *Environment and the Law. An Introduction for Environmental Scientists and Lawyers*. Longman, 1996.

McEwen, John *Who Owns Scotland?* EUSPB, 1977.

McIlvanney, William *The Big Man*. Hodder and Stoughton, 1985.

McKay, Ron *Mean City*. Hodder and Stoughton, 1995.

Maclean, Alasdair *Night Falls on Ardnamurchan*. Victor Gollancz, 1984.

McLean, Duncan *Blackden*. Secker & Warburg, 1994.

MacLellan, Rory 'Tourism and Scotland's Environment', in *Tourism in Scotland* by L. R. MacLellan and R. Smith (eds.). Tourism and Hospitality Series, Scottish Tourism Research Unit, Thompson Business Press, 1998.

McWilliam, Candia *A Case of Knives*. Bloomsbury, 1988.

Meacham, Standish *Regaining Paradise. Englishness and the Early Garden City Movement*. Yale University Press, 1999.

Meadows, D. H., Meadows, D. C., Randers, Jørgen and Behrens, William W. III *The Limits to Growth. A Report for the Club of Rome's Project on the Predicament of Mankind*. Earthscan, 1972.

Meadows, D. H., Meadows, D. C., Randers, Jørgen and Behrens, William W. III *Beyond the Limits to Growth. Global Collapse or a Sustainable Future*. Earthscan Publications, 1992.

Mitchell, Ian *Isles of the West. A Hebridean Voyage*. Canongate, 1999.

Mitchell, John *Loch Lomondside: Gateway to the Western Highlands of Scotland*. HarperCollins, 2001.

Monbiot, George *Captive State: The Corporate Takeover of Britain.* Macmillan, 2000.

Mooney, Gerry 'Changing Places: Perspectives on the Development of a Municipal Suburban Landscape', in *Images of the Street: Planning, Identity and Control in Public Space* by Nicholas R. Fyfe (ed.). Routledge, 1998.

Moscovici, Serge 'The phenomenon of social representations', in *Social Representations*, by R. M. Farr and S. Moscovici (eds.). Cambridge University Press, 1984.

MVRDV *Metacity Datatown*. 010 Publishers, 1999.

Naess, Arne *Ecology, Community and Lifestyle: Outline of an Ecosophy.* Cambridge University Press, 1990.

Oelschlaeger, Max *The Idea of Wilderness. From Prehistory to the Age of Ecology*. Yale University Press, 1991.

O'Hagan, Andrew *Our Fathers.* Faber and Faber, 1999.

O'Hagan, Andrew 'The End of British Farming' *London Review of Books* 23(6): 3–16, 2001.

O'Rourke, Daniel *Dream State: The New Scottish Poets*. Polygon, 1994.

Orwell, George *The Lion and the Unicorn*. Penguin Books, 1941.

Orwell, George *The English People*. Collins, 1947.

Page, David 'The Censorship of Neglect: Morris & Steedman's Private Houses', in *Rebuilding Scotland: the Postwar Vision 1945–1975*, by Miles Glendinning (ed.). Tuckwell Press, 1997.

Paterson, Anna (ed.) Introduction, in 'The Environment in Contemporary Swedish Writing'. *Swedish Book Review Supplement*, 1997.

Patterson, I. A. P., Reid, R. J., Wilson, B., Grellier, K., Ross, H. M., Thompson, P. M. 'Evidence for Infanticide in Bottlenose Dolphins: an Explanation for Violent Interactions with Harbour Porpoises?' Proceedings of the Royal Society B, 265: 1167–70, 1998.

Paxman, Jeremy *The English. A Portrait of a People*. Penguin Books, 1999.

Peat, Jeremy and Boyle, Stephen (eds.), with Bill Jamieson *An Illustrated Guide to the Scottish Economy*. Duckworth, 1999.

Pfoffenberger, M. *A Long-term Strategy of the IUCN Working Group on Community Involvement in Forest Management*. IUCN, 1996.

Planterose, Bernard *Report on Timber Specification for the National Trusts Scotland Visitor Centre in Glencoe*. North Woods Construction Ltd, October 1999.

Power, William *Literature and Oatmeal. What Literature has meant to Scotland* in the series *The Voice of Scotland* by Hugh MacDiarmid (ed.). George Routledge & Sons, 1935.

Rankin, Ian *Black and Blue*. Orion, 2000.

Rankin, Ian *Set in Darkness*. Orion, 2000.

Reid, Alastair 'Weathering', a poem from 1978, quoted in *Scotland – An Anthology* by Douglas Dunn. HarperCollins, 1991.

Robertson, James *Scottish Ghost Stories*. Warner Books, 1996.

Robertson, James *The Fanatic*. Fourth Estate, 2000.

Robinson, Nicholas A. (ed.) *Agenda 21: Earth's Action Plan*. Oceana Publications, 1993.

Rogers, Richard *Towards an Urban Renaissance*. Reported by the Urban Taskforce led by Lord Rogers of Riverside to the UK Department of the Environment, Transport & the Regions, 1999.

Rogers, Richard and Gumuchdjian, Philip *Cities for a Small Planet*. Faber and Faber, 1997.

Rose, Chris *The Dirty Man of Europe – The Great British Pollution Scandal*. Simon & Schuster, 1990.

Rowell, Andrew *Green Backlash. Global Subversion of the Environment Movement*. Routledge, 1996.

Schama, Simon *Landscape and Memory*, HarperCollins, 1995.

Schumacher, Ernest *Small is Beautiful*, Abacus, 1974.

Scott, Andrew Murray *Tumulus*. Polygon, 2000.

Scott, Michael and Scott, Sue (eds.) 'Conservation News: Cairngorms National Park', in *Scottish Environment News* [SCENES], February 2001.

Scott, Walter *The Lay of the Last Minstrel*, 1805. Printed for Longman, Hurst Rees and Ormes by James Ballantyne & Co, 1810.

Scruton, Roger *England: an Elegy*. Chatto & Windus, 2000.

Slesser, Malcolm *The Politics of the Environment. A Guide to Scottish Thought and Action*. George Allen & Unwin, 1972.

Smout, T. C. *Scotland since Prehistory. Natural Change and Human Impact*. Scottish Cultural Press, 1993.

Smout, T. C. *A Century of the Scottish People 1830–1950*. Fontana Press, 1997.

Smout, T. C. *A History of the Scottish People 1560–1830*. Fontana Press, 1998.

Smout, T. C. and Lambert, R. A. (eds.) *Rothiemurchus. Nature and People on a Highland Estate 1500–2000*. Scottish Cultural Press, 1999.

Smout, T. C. *Nature Contested. Environmental History in Scotland and Northern England since 1600*. Edinburgh University Press, 2000.

Smout, T. C. (ed.) 'Nature, Landscape and People since the Second World War'. A conference held at the Royal Society of Edinburgh. Tuckwell Press, 2001.

Spence, Alan *Its Colours They Are Fine*. The Salamander Press, 1983.

Spence, Alan *Way to Go*. Phoenix House, 1998.

Stokols, D. and Altman, Irwin (eds.) *Handbook of Environmental Psychology*. Wiley, 1987.

Taylor, Nigel *Urban Planning Theory since 1945*. Sage Publications, 1998.

Thomas, M. F. and Bibby, J. S. (eds.) *Evaluation of Land Resources in Scotland*. Proceedings of the Royal Geographical Society Symposium, University of Stirling, 1989.

Thomas, M. F. 'Past and Future Decades of Land Evaluation in Scotland', in *Evaluation of Land Resources in Scotland* by J. S. Bibby and M. F. Thomas (eds.). Proceedings of the Royal Geographical Society Symposium, University of Stirling, 1989.

Thompson, P. M., Wilson, B., Grellier, K., Hammond, P. S. 'Combining Power Analysis and Population Viability Analysis to Compare Traditional and Precautionary Approaches to the Conservation of Coastal Cetaceans'. *Conservation Biology* 14: 1253–63, 2000.

Thorén, Roger (ed.) 'On Environmental Technology' ('Tema Miljöteknik'), in *Miljö Rapporten: Ekonomi + Teknik*, April 2000.

Torrington, Jeff *Swing Hammer Swing!* Secker & Warburg, 1992.

Torrington, Jeff *The Devil's Carousel*. Secker & Warburg, 1996.

Wallace, Gavin and Stevenson, Randall (eds.) *The Scotttish Novel since the Seventies: New Visions, Old Dreams*. Edinburgh University Press, 1993.

Warner, Alan *These Demented Lands*. Jonathan Cape, 1997.

Warren, Stacy '"This Heaven Gives Me Migraines". The Problems and Promises of Landscapes of Leisure', in *Place/Culture/Representation* by James Duncan and David Ley. Routledge, 1993.

Watson, R. T., Noble, I. R., Bolin, B., Ravindranath, N. H., Verardo, D. J., Dokken, D. J. (eds.) *Land-Use, Land-Use Change and Forestry*. IPCC Special Report, 2000.

Welsh, Irvine *Trainspotting*. Secker & Warburg, 1993.

Wigan, Michael *The Scottish Highland Estate. Preserving an Environment*. Swan Hill Press, 1991.

Wightman, Andy *Who Owns Scotland*. Canongate, 1996.

Wightman, Andy *Scotland: Land and Power*. Luath Press, 1999.

Wightman, Andy 'Land Reform: Politics, Power and the Public Interest' The 1999 McEwen lecture. www.caledonia.org.uk/land/ lectures

Wilson, Norman (ed.) *Scottish Writing and Writers. A Survey of Modern Scottish Literature*. Ramsey Head Press, 1977.

Wilson, Peter 'A Manifesto for an Architecture Policy for Scotland', in *The Case for Change: an Architecture Policy for Scotland*. ARCA 1, 1999.

Wood, Emma *Notes from the North. Incorporating a Brief History of the Scots and English*. Luath Press, 1998.

Wood, Michael *In Search of England. Journeys into the English Past*. Penguin, 2000.

Index

Note: all page references to illustrations are given in **bold**.

Aberdeen, 22, 27, 101, 158, 177, 215, 248
Aberfeldy, 169
Abernethy Forest, 139
Abriachan, Loch Ness, 232
'access', to countryside, 111, 115–16, 135, 140, 143, 146, 149
see also nature conservation
Advance Brazil, 225
Africa, 27, 103
Agenda 21, 8, 15–16n, 51, 66, 88
Agenda 2000, 215, 216
Alcan, 182
Amazon rainforest, 225
American writing, 38
Angus, 85, 131–2, 234, 236
Animal Farm, 19
animal welfare, 79, 100–1
'architectural laziness', 187–8
architecture
and environment, 159–88
animal architecture, 171
Manifesto group, 165
see also cities, planning
Ardnamurchan, 23
Argentina, 100
Argyll, Argyllshire, 21–2, **246**
Arlanda Airport, 135–6
Arnold, Matthew, 161
Asia, 103
Asia-Pacific Economic Corporation, 220–1
Association for the Protection of Rural Scotland, 56
Association of Regional and Island Archaeologists, 56
Association for Scottish Ski Areas, 57
Assynt, 150, 239
Athanasiou, Thomas, 7–8
Atkinson, Kate, 27
Atomic Energy Authority, 61
Austria, 198
Aviemore, 131
Partnership, 59, 130
Axelsson, Ake, 208–9
Ayrshire, 52, 115

Badenoch and Strathspey Conservation Group, 56, 107
Balloch, Loch Lomond, **127**
Balloch Castle Country Park, 129
Baltic Sea, 204, 205
Bank of America, 249
Banks, Iain, 28
Bannister, Jon, 178
Barcelona,180
Barnes, Julian, 33
Barthes, Roland, 31
Battle for Scotland, The, 37
Bede, 32
Begg, David, 248
Beinn a'Bhuird, **99**
Belgium, 50
Bell, Ian, 54
Bellamy, David, 103
Ben Lui, 128,
Ben Nevis, 139
Bernera, Isle of Lewis, **252**
Best Practicable Environmental Options (BPEOs), 191, 247
Beuys, Joseph, 223
Big Man, The, 25–6
Biological Recording in Scotland, 56
biotechnology, 137, 250
Birmingham, 163
Bisset, Jack, 127–8
Black Cuillin, Skye, 149
Blackden, 22
Blair, Tony, 54
Blueprint for Survival, 7
Bodmin Moor, 218
Borders region, 20–1, 115–16, 195, 232–3

Index

Borgie Forest, 231
Botanical Society of Scotland, 57, 59
Boyd, William, 18, 27
Boyle, Stephen, 192–3
BPEOs *see* Best Practicable Environmental Options
Braer oil spill, 9–10, 247–8
Brand, Janet, 168
Brazil, 100
Brazzaville Beach, 27
Brent Spar, 247
Bridge, The, 28
Britain, Britishness, 33, 35, 39–40
British and International Greenkeepers Association (BIGGA), 113
British Association of Nature Conservationists Scotland, 56
British Association of Settlements and Social Action Centres (BASSAC), 62
British Association for Shooting and Conservation, 57, 142
British Nuclear Fuels, 78
British Trust for Conservation Volunteers (BTCV Scotland), 56
British Trust for Ornithology, 56, 142
British Urban Regeneration Association, 185
British Wind Energy Association, 61
Broadhurst, Richard, 227–8
Browne, Anthony, 101
Brownie of Bodsbeck, The, 44
brownies, 44
Brundtland Commission, 8
BSE disaster, 12, 54,
Building Scotland, 166
Burghead, 66
Butterflies and Moths of Europe, The, 21
Butterfly Conservation Scotland, 56, 113
Byatt, A. S., 20

CADISPA, 65–8, 219
CAP *see* Common Agricultural Policy
Cairngorms, Cairn Gorm, **14**, 58–60, 70–1, 107–8, 115–16, 117, 125, 129–34, 233
  Campaign, 56, 70–1
  funicular railway, 70–1, 107, 116, 130–4
  National Park, 85, 107, 133–4
  Partnership, 58–60, 84–5, 238
Caithness, 22
  Caithness and Sutherland Enterprise, 64
Caledonian Forest, 133, 226, 227
Callander, **110**
Callander, Robin, 145, 149–50
Callanish, 105
Calmer Waters, 111, 112
Cambuslang, 21–2, 72
Campaign for Environmentally Responsible Tourism, 97
Canada, 198
Captive State, 78, 249
carbon dioxide emissions, 159–60, 191–2, 225, 226
  forests in controlling, 223–6, 241n
Carlsson, Lars, 232
Castells, Manuel, 163
Centre for Alternative Technology, Wales, 222
Centre for the Analysis and Demonstration of Demonstrated Energy Technologies, 60–1
Chernobyl, 9
chromium contamination, 70, 72, 75n
Cities for a Small Planet, 166
cities, cityscapes, vii, ix, 11, 157–88, 206, 253, 254
  in literature, 19–20, 23–8, 41–4
  see also urban regeneration
climate change, 51
  see also environment, forests
Clone City, 159–61, 163
Clyde, River, **25**, 72, 117, 178
  Clyde Valley Regional Plan, 176
Coigach, 150
Coll, 141
Colonsay, 66, 67–8
Colorado, 103
Common Agricultural Policy (CAP), 123, 215, 216
Common Fisheries Policy, 96
Common Ground, The, 6
communities, community projects, 59, 60–72, 78–81, 84–5, 90–2, 219–23, 230–3
Community Councils, 59
Community Development Centres, 62
Community Development Foundation, 62
Community Matters, 62
Community Self-Build Scotland, 165
Community Work Training Groups, 62
conservation *see* nature conservation
Conservative government and Party, 12, 37, 139–40, 249
Convention of Scottish Local Authorities (CoSLA), 51, 57
Cornwall, 32, 103, 163
Corporate Scotland 1998/9, 8–9

Costa Brava, 117
Council For Scottish Archaeology, 56
Council for British Industry Scotland, 215
Countryside Acts, 5, 143
Countyside Alliance, 147
Countryside Code, 143
Countryside Commission, 7
countryside crisis, 13
Countryside Premium, 238
Craig, Cairns, 18–19, 21, 26
Cramb, Auslan, 245, 248
Cranhill Water Tower, 180
Crawford, Robert, 37–8, 42
Crianlarich, **22**
Crofters Commission, 64
crofting, 22, **64**, 79, 230, 232–3
Crofts, Roger, 124, 137–9
Cruickshank, Geri, 174–5
Cuba, 103
Culture and Anarchy, 161
Curry-Lindahl, Kai, 132

Dalhousie Estates, 236, 238–9
Dalhousie, James, 236–9
Dark Horse, 43
Darling, Frank Fraser, 6–7
Dartmoor, 218
Death of England, The, 36
Death of the Village in Late 20th Century Europe, The, 216
'deep ecology', 5
Deeside Forest, 238
deforestation, x–xi
  see also forests
democracy, vii, viii, 11, 57–8, 121, 254
Denmark, 39, 73, 136, 173, 198
Department of the Environment, Transport and the Regions, 161, 166, 194, 225
Developments Trust Association, 61
Devil's Carousel, The, 26
devolution, in Scotland, vii, 7
Dewar, Donald, 8–9, 12, 138
Dirty Man of Europe, The, 6
Disaffection, A, 26
dolphins, 100–1
Doric language, 18
Down to Earth, 83
Dream State, 37
Dunbarton Business Education Project, 129
Dunblane, 182
Duncan, James, 31
Dundee, 24, 27–8, 158, 175–6
  District Council, 176
Dunion, Kevin, viii, 68–74, 191–2
Dunn, Douglas, 20, 45
Dunnet, Dorothy, 29–30
Dunsire, Ian, 167–8
Dùthchas, 63–5, 67, 219
Duthie, Niall, 221
Dyfi Valley, Wales, 223–4

Earth Summit, Rio, 8, 66, 88
Earth, Wind and Water, 111
Easdale, 66, 67–8
East Anglia, 36
East of Scotland Water, 172, 173, 195
Ecological Party, 6
  see also Green Party
ecology, concept of, 4–5, 9, 36, 51, 80, 201
  and consumers, 77
  'eco-city', 162–4
  eco-houses, 168–9
  eco-manufacturing, 198–201, 205, 206–9
  eco-tourism, 96–108
  eco-village movement, 61
ECOTEC Research and Consulting Ltd, 220–1
Eden Project, 103
Edinburgh, 9, 24, 26–7, 29, 30, 44, 73, 94, 148, 158, 164, 167, 168 220, 228
  Zoo, 171
Education 21 Scotland Forum, 88
Edwards, Robert, 103, 132, 229, 248
Eigg, 141, 148, 150, 151
Ekman, Kerstin, ix, x–xi
Elephtheriou, Margaret, 247
  Tasso, 247
Elkington, John, 192, 201
Elphinstone, Margaret, 29
Emotionally Weird, 27
Energy Action Grants Agency, 61
Energy Question, The, 6
Energy Support Unit Centre for Alternative Technology, 60
Engineer in the 21st Century, The, 86
England, English, 5, 69, 86,161, 163, 168, 173, 217–18, 226, 250, 254
  identity 31–4, 39–40
  see also British
England: An Elegy, 33
England, England, 33
English People, The, 33
environment, environmentalism, vii, viii, 1, 4–13, 50–4, 146, 192–4, 201
  and consumers, 77, 91–2
  and education, 62, 65–8, 77–92, 162, 221–2

and landscape, *see* landscape
and law, 11, 72, 234, 235
see also ecology, Green Party, land reform, National Parks, planning, pollution
Environment Agency (England and Wales), 91
Environment and Planning, 193
Environment Group, 195–6
environmental history, 2–3, 5–6, 139
Environmental Impact Assessment (EIA), 201, 247
Environmental Impact Legislation, 234, 235
Environmentally Sensitive Areas, 127, 238
Essential Scotland, 12
European Action Plan on the Environment, 85
European Investment Bank, 249
European LIFE Programme, 63
European (Nature) Conservation Year 1970, 5
European Regional Development Fund, 129
European Round Table of Industrialists, 78
European Round Table on Cleaner Production, 196–7
European Structural Fund, 130,
European Union, 66, 67, 69, 71, 79, 80, 96, 101, 131, 136, 174, 196–7, 199, 204, 213, 216, 217, 234
Eco-management and Audit Scheme (EMAS), 197
Rural Development Policy Programme, 216
Everything You Need, 29
Exmoor, 218
Extended Producer Responsibility (EPR), 199–200

Fagan, Geoff, 65–8
Fallow, Jeff, 36
Fanatic, The, 30, 42
Farm Woodland Premium, 238
farming, 3, 59, 79–80, 91, 138, 141, 146, 213–14, 250
see also organic farming

Fife, 27, 182
Regional Council, 167, 172–3
Findhorn Foundation, 61
Finland, 136–7, 198, 204, 215, 216
Firth of Forth, 29, 139, 234
fish-farming, 13, 73, 218, 245–7, 256n
fisheries, 245–7
Flanders Moss, 126
Flow Country, 231
Flying Scot, 57
Food Standards Agency, 54, 73
forest, forestry, x–xi, 59–60, 146, 218, 223–39, 254
and climate change, 223–6
ownership, 226, 231–3, 234–5, 237
Forest Enterprise, 64, 226, 231, 236, 237
Forest Industries Development Council, 229–30
Forestry Commission, 64, 226–37, 238
Conservators, 234
Forres, 182
Forum for the Future, 86
Forum on the Environment, 6
Forward Scotland, 51, 65, 67, 80, 172, 183, 194
Foundation for Strategic Environmental Research (MISTRA) 203, 204, 205
Fragile Land: Scotland's Environment, 7, 245, 248
France, 173
Freedom-to-Roam Campaign, 143–4
Friends of Loch Lomond, 56
Friends of the Earth Scotland, viii, 54, 68–73
Fuller, Buckminster, 163
Fyfe, Nicholas, 178

Gaelic language, 18
writing, 42
Gaia Group, 110, 167, 168–71, 239
Garrick Committee, 90
Gartmore, 235
Geddes, Patrick, 36, 158
genetically modified organisms, 13, 72–3
Germany, 6, 8, 168, 172, 182, 196, 201
Giddens, Antony, 58
Gilbert, John, 182–3
Glancey, Jonathan, 166
Glasgow, 6, 21–2, 24–6, 29, 41, 43, 44, 94, 117, 125, 129, 158, 159, 162, 164, 167, 168, 220, 234, 248
crime, 178–9
Herald, 54, 102; *Sunday Herald*, 103, 177, 229, 248
housing, 176–9
Housing Associations, 168, 181–4
Partnership, 180
Year of Architecture and Design, 62, 179–81, 185, 206
Glen Feshie, 130

Glencoe, 117
Visitor Centre, 169, 239
Glendinning, Miles, 159–62
Glenlivet Estate, 85
Glenrothes, Fife, 160
Gliding Association, 57
global resource consumption, 192–4
globalisation, vii, 31–2, 35, 77–8
golf, environmental aspects, 112–14
Gordon of Strathblane, James, 102
Govan, **25**
Governance of Scotland Forum, The, 34–5
Grafton Water, 143
Grampian Forest, 238
Grantown on Spey, 85
Gray, Alasdair, 1–2, 18–19, 24
Greece, 216
Green Backlash, 8
Green Capitalists, The, 6, 192
Green movement, 4
Green Party, 54
Scotland, 6, 12, 78–81, 254
Green Tourism Business Award Scheme, 104
green belt, 108–9, 163
'green debate' *see* environment, Green Party
Greening the Design Curriculum, 86
Greenpeace, 57, 247
Greenwich, London, 163
Greig, Andrew, 20–21
Grieve, Christopher, *see* MacDiarmid
Grieve, Robert, 140
Gunn, John, 168, 174
Gunn, Neil, **22**, 43
Gust of Wind, A, ix

Habitat II, 15
Halliday, Sandy, 168, 171
Harper, Robin, 78–81
Harris, 10, 70, 148, 219, 251
Harvey, Graham, 215
Harvie, Christopher, 35–6
Hawaii, 117
Hawken, Paul, 70, 75n
Heather Trust, 57
Hebridean Whale And Dolphin Trust, 56, 102
Heijne, R., 187
Heisenberg Group, 179
Helmsdale, 219
Heriot-Watt University, 195, 222
Centre for Environmental Resource Management, 195
Centre for Island Technology, 222
Hertz, Noreena, 77, 250
HI-Ways, 63–4, 221
Highland and Islands, 43, 58, 63–4, 81, 115, 117, 130, 215, 220, 230–3
Highland and Islands Enterprise, 59, 63–4, 131, 132, 220, 221, 230, 231
Highland and Islands ICT Network, 221–2
Highland Clearances, 218
Highlands and Islands, University of the, 195, 221–3
Highlands Birchwoods, 57
Highlands of Scotland Tourist Board, 64
Historic Scotland, 54, 59, 64, 67, 114
Hobsbaum, Philip, 43
Hogg, James, 42, 44
Home Energy Efficiency Scheme, 61
House, Syd, 233–6
Housing Association, 172, 181–5
How Late it Was, How Late, 26
Howard, Ebenezer, 176–7
Humanities Research Council, 203
Hunter, James, 145, 231
Hutton, Will, 163

India, 200
Indonesia, 9
Industrial Revolution, 11
Ingold, Tim, 31, 151–2
Inman, Philip, 54
Institute of Offshore Engineering, 195
Intergovernmental Panel on Climate Change, 223–4
International Centre for Island Technology, 195
International Institute for Applied systems Analysis, 201–2
International Institute for Industrial Environmental Economics, 198, 199, 200
Inverness, 24
Ireland, 247
Irvine, 161
Islay, 148
Isles of the West, 147–8
Its Colours They Are Fine, 24

Jack, Ian, 177
Jacobs, Jane, 178–9
Japan, 114, 202
Jarrow, Northumberland, 32
March, 32
Jenkins, Robin, 20, 21–2
Jensen Foundation, 235
Jernelöv, Arne, 201–2, 204
John Muir Trust, 56, 139, 147–8

Jorwerd, 216–17
Journal of Environmental Planning and Management, 193

Kailyard School, 43
Kelman, James, 26, 43–4
Kelvingrove Museum, Glasgow, 181
Kennedy, A. L., 19, 29, 39–42, 44
Kenya, 103
Killing of the Countryside, The, 215
King Hereafter, 29
Kingask Development, St Andrews, 114
Kingdom Housing Association, 172
Kinross, 234
Kirkcaldy, 173
Kiruna, Sweden, 121
Klein, Naomi, 77–8, 250
Knoydart, 150, 151
Kuppner, Frank, 29
Kyle of Tongue, **152**

Labour Party, 12, 55
LaFarge Redland, 71
Laggan Forest, 231, 232–3
Lake Windermere, 127
Lanark, 1–2, 18–19
Lanarkshire, 70, 72
Land Fund, 150, 219–20, 231
land ownership, 71–2, 94, 107, 121, 130, 144–52, 238
 in Sweden, 134–5
land reform, 13, 62–3, 137, 144–52
 Act, 63, 94–5, 144–6, 214
landfill tax credits, 80, 194
landscape,
 assessment, 49–50, 250–3
 attitudes to, vii, x–xi, 1, 3, 8–12, 17, 21–3, 30–1, 41–2, 43–5, 49–50, 73, 129–30, 244–55
 and National Parks, *see* National Parks
 and Scottish identity, *see* Scottish identity
Landscape Institute, 251
Langholm, Borders region, 43
Lantra Trust, 57
Laslett, Peter, 33
LEADER, LEADER II *see* European Union Rural Development Policy Programme
Learning for a Sustainable Future, 4–5
Learning for Life, 82–3, 85, 87, 89
Legge, Gordon, 29
Leicester, Graham, 12
Leith, 175, 219
Leopold, Aldo, 145, 155n
Leven, 219
 River, 126, 128, 129
Leverhulme, William, 219
Lewis, Outer Hebrides, 252
Lewis, Penny, 166
Ley, David, 31
Liddell, Howard, 167–71
Life Cycle Analysis, *see* sustainable development
Limits to Growth, 7
Lindhquist, Thomas, 199–201
Lingarbay, 70–71
Lion and the Unicorn, The, 33
liquid petroleum gas, 116, 120n, 127, 215
literature, *see* American, Scottish, Swedish writing
Lloyd, John, 12
Lobster Moth, 21
Local Enterprise Companies, 67, 105, 106, 107
Locate in Scotland, 71
Loch Katrine, 176
Loch Lomond, 115–16, 124–9, **127**, 134, 234
 Shores project, 129, 130
Loch Lomondside, 128–9
Loch Morlich, **14**
Loch Ruskie, 126
Loch Slapin, 23
Lochawe Meadows, 144
Lochnagar Hills, 239
London Design Museum, 163
London Review of Books, 213
Long, H. K., 24–5
Los Angeles, 160
Lovin, Amory, 192, 196
Lumphinnans, 183
Lund University, Sweden, 198, 199–201
Lynch, Michael, 35

MacArthur, Alexander, 24–5
MacDiarmid, Hugh, *pseud* Christopher Grieve, 27, 43
McCrone, David, 3
McEwen, John, 145
McFadden, Jim, 168
McIlvanney, William, 25–6
McKay, Ron, 25
McLaren Hall, Callander, **110**
Maclean, Alasdair, 23
McLean, Duncan, 22
McLellan, Rory, 105–8, 124
McLeod, John, 149
McLetchie, David, 12
McWilliam, Candia, 17

Major, John, 40
Mar Estate, 130, **133**, 254, **255**
Marine Conservation Society, 56
Marr, Andrew, 37
Maureen Alpha, 247
Maxwell, Gavin, 22–3
May, Isle of, 142
Mean City, 25
Melness, **64**
Metacity Datatown, 159–62, 224–5
Methil, 172–3
Mexico City, 158
Middleton, Fraser, 167, 171–3
Millennium Commission Funding, 222
Dome, 163
Forest for Scotland, 227, 229–30
Lottery Fund, 67
Ministerial Group on Sustainable Scotland, 68
Ministry of Agriculture, Food and Fisheries (MAFF), 213, 215
Ministry for the Environment, Scotland, 58–9
Ministry of the Environment, Transport and Regions, 54
Ministry for Rural Development, Scotland, 58
Mitchell, Ian, 23, 55, 56, 147–8
Mitchell, John, 128–9
Mitchison, Naomi, 43
Monbiot, George, 78, 249, 250
Moray, 131–2
Firth, 61, 101; Moray Firth Partnership, 63
Morgan, Edwin, 37
Morrison plc, 131–2, 164
Mountain Rescue Association, 57
Mountaineering Bothies Association, 59
Mountaineering Council of Scotland, 56
Mugdrum Island, 142
Muir, John, 107
Mulholland, John, 103
Munn and Dunning report, 87
Munn, James, 87
Murdoch, Iris, 20
MVRDV group, architects and planners, 159–62, **197**

Naess, Arne, 5
National Botanical Gardens, Wales, 222
National Energy Action, 61
National Environmental Technology Centre, 61
National Lottery, 61, 67, 114
Charities Board, 67
funding, 54, 67
National Nature Reserves, 126, 128
National Parks, 5, 13, 58, 107–8, 117, 121–34, 138, 140–4, 150, 214, 234, 238
Act (Scotland) 2000, 124–5, 137, 141
in Sweden, 134–7, 144, 146
National Planning Guidelines, 186–7
Regulations, 172–3
National Trust for Scotland, 56, 147–8, 169–70
Natura 2000, 136, 154n, 195–6
Natural Step, The, 70, 74
Nature Conservancy Council, 7
nature conservation, 4–10, 95–6
and access to countryside, 96–114, 116–17, 121–44
and land reform, 144–52
see also environment, landscape
Nature Contested, 7, 124
Nature Reserves, 143
Nature, Landscape and the People Since the Second World War, 5
Netherlands, 6, 8, 73, 86, 143, 158–9, 182, 196, **207**, 216–17, 254
Design Institute, 187, 206–9
Network for Alternative Technology and Technology Assessment, 61
Neutelings, Willem Jan, 187–8
New and Renewable Energy Programme, 61
New Lanark, 185–**6**
New Opportunities Funding, 194
New Zealand, 100
No Logo, 77–8
No Mean City, 24–5
non-governmental organisations, 7, 54, 56–60, 68–74, 124, 132, 141, 222
see also quango, United Nations
Nordic Council, 136
North East Mountain Trust, 56
North Of Scotland Water Authority, 64
North Queensferry, 52
North Sea, 247
Northern Constabulary, 63–4
Norway, 63, 101, 136–7, 168, 198, 215, 248

O'Hagan, Andrew, 18, 26, 27, 40–1, 213–14
O'Rourke, Danny, 37–8
Observer (Scotland), 103
oil exploration, 245, 247–8
Olsson, Mats, 134–7
Open College Network, 62
Oresund Straits, 203–4

Organic Aid, 238
organic farming, organic foods, 77, 79–80, 196, 213, 238, 254
Organisation of Economic Co-operation and Development, 191
Orkney Islands, 195, 222
Orwell, George, *pseud* Eric Blair, 19–20, 33
Osborne, Michael, 236
Our Fathers, 26, 177
Overseas Development Institute, 232
Owen, Robert, 185–6

Page, David, 159–61, 184–5
Paisley, 170
Parthenon, 166
'Passivhaus', 172
Patagonia, 208
Paxman, Jeremy, 32, 33
Peabody Trust, 163–4
Pearce, David, 201
peat conservation, 126
Peat, Jeremy, 192–3
Peer Gynt, 44
People's Centre for Renewable Energy, Copenhagen, 198–9
Perception of the Environment, The, 31, 151–2
Perth, Perthshire, 131–2, 164, 168–9, 234
  Fairfield Co-operative, **169**
Place/Culture/Representation, 31
planning, planning law, 52–3, 70, 123–4, 137, 158, 159–62, 163, 164, 165–6, 175–6, 176–8, 184–7
Planterose, Bernard, 239
Plantlife, 56
Pleasure in Scottish Trees, A, 232
Politics of the Environment, The, 6
pollution, 3, 5, 6, 72, 51, 83–4, 91, 101, 116, 123, 128, 134, 195–6, 204
  and human health, 55
  legislation, 6
Poor Angus, 20–2
Popular Flying Association, 57
population, 161, 185, 217
Population Bomb, The, 7
Power of Identity, The, 163
Private Eye, 147
Private Finance Initiative (PFI), 164

quangos, 2, 15n, 54, 55–74, 84–6, 88–9, 109, 125–7,147–8, 202, 221–2
  scientists as members, 140

Ramblers' Association Scotland, 56, 111, 131
Rankin, Ian, 26–7
recycling, 3, 6, 67–8, 80, 182, 186–7, 196, 198–9, 200, 208
  see also architecture, planning
Reforesting Scotland, 56
Regeneration of Scotland Award, 166
renewable energy, 60–1, 73, 223, 226
Renfrew, 170
Rhum, 148
Rifkind, Malcolm, 12, 139–40, 155n
Robertson, James, 30, 42–5, 47n
Rocky Mountain Institute, 196
Rodel, 73, 251–3
Rogers, Richard, 166
Roineabhal Mountain, 70–1
Rose, Chris, 6
Ross-shire, 36
Rothiemurchus Estate, 130, 146
Rowell, Andrew, 8
Royal Forestry Society, 230
Royal Geographical Society, 57
Royal Incorporation of Architects in Scotland, 166
Royal Scottish Geographical Society, 57
Royal Society for the Protection of Birds (RSPB), 56, 89, 106, 130, 139, 141, 147–8
Royal Zoological Society, 88
Rural Affairs Department (Scotland), 64
Rural Forum Scotland, 59
Rural Stewardship Scheme, 216
'rural development', 94–6, 141, 145, 186–7, 213–39
  and community projects, 219–23
  and farming, 213–19
  and forests, 223–39
  and public transport, 215
  Regulation, 216
Russia, 200
Rutherglen, 72
Rutland Water, 144

SafeinHerit Network, 63
St Andrews, 107, 114, 139
Salmond, Alex, 12, 55
Saltire Society, 56
Sandaig, 22
Sandford Principle, *see* nature conservation
Scandinavia, 8, 84, 143, 157, 166, 196
  writing, 41–2
Scotland – the Brand, 3, 35, 117
Scotland for Beginners, 36
Scotland, Scottish,
  history, 1, 18, 29–34, 147

housing, 161–6, 168–86
identity, vii, 1, 2, 17–18, 22–3, 33–8, 40–4
land ownership, 71–2, 94, 107, 121, 130, 144–52, 238
writing, 11, 17–45, 94
Scotland's Environment: the Future, 7
Scotland: Land and Power, 7
Scotland: the Challenge of Devolution, 191
Scots language, 18
Scott, Andrew Murray, 28
Scott, Sir Walter, 28, 30, 94, 251
Scottish Arts Council, 59, 64
Scottish Centre for Art Extraordinary, 174–5
Scottish Council for National Parks, 57
Scottish Country Landowners Federation, 57
Scottish Countryside Activities Council, 57
Scottish Countryside Rangers Association, 57
Scottish Crofters' Union, 230
Scottish Ecological Design Association, 168, 170, 171
Scottish Education Bill (2000), 90
Scottish Enterprise, 59, 71, 129, 194–5, 219–20, 221, 229, 250
Scottish Environment LINK, 56, 59, 124, 128, 141
Scottish Environment Protection Agency (SEPA), 7, 64, 69, 74, 80–1, 83, 86–7, 88–9, 90–2, 95, 96–8, 113, 115, 116, 128, 172
Scottish Environmental Design Association, 89
Scottish Environmental Education Council, 82, 88, 89
Scottish Executive, viii, 51, 52, 55, 56, 59, 137–8, 145, 181, 214–16, 217, 249
architectural policy, 165–6, 177
Development Department, 52
environmental policy, 53–4, 58–9, 62, 63, 67, 69, 71, 72, 87, 90, 133, 168, 170, 191–2, 194, 216, 225–30, 236–7, 248
Rural Affairs Department, 67, 195–6, 214
Scottish Field Studies Association, 57
Scottish Forest Industries Cluster, 229
Scottish Forest Strategy, 226–30, 234–5, 236
Scottish Golf Wildlife Initiative, 113
Scottish Higher Education Funding Council, 86
Scottish Highland Estate, The, 146
Scottish Homes, 59, 64, 164, 166, 175, 181, 182, 183, 184
Scottish Institute for Sustainable Technology, 195, 210n
Scottish Landowners Federation, 59, 140, 142–3, 146, 150, 236
Scottish Mountaineering Club, 57
Scottish Mountaineering Council, 85
Scottish National Farmers Union, 57
Scottish National Party, 55
Scottish Native Woods, 57
Scottish Natural Heritage, 7, 53, 67, 80–1, 90, 106, 107, 108, 114, 116, 125, 130, 132, 133–4, 137–40, 142, 143, 239
Scottish Office, 2, 49, 59, 80, 82, 159, 249
Scottish Ornithologists Club, 57
Scottish Parliament, 3, 12, 36, 55, 78–80, 96, 106–7, 143–4, 236–7, 245–6
Environment Committee, 72
'Scottish Renaissance', 22–3
Scottish Scenic Trust, 57
Scottish Sports Council, 107, 109
see also SportScotland
Scottish Tourism and Environment Forum, 101, 104
Scottish Tourism Co-ordinating Group, 105
Scottish Tourist Board, 64, 98–9, 102, 105, 117
see also VisitScotland.
Scottish Wild Land Group, 57
Scottish Wildlife Trust, 57, 126, 141
Scottish Woodlands Ltd, 236
Scottish Writers and Writing, 44
Scruton, Roger, 32, 33, 39
sea, problems of, 244–8
Seafield Estates, 237
Secretary of State for Scotland's Advisory Group on the Environment, 68
Set in Darkness, 26–7
Sewel, Brian, 82
Shama, Simon, 223
Sheridan Group, 129
Shetland Islands, 9–10, 71, 247–8
Shetland Wildlife Holidays, 100
Shoe, The, 29
Silent Spring, The, 7
Silicon Glen, 38
Sites of Special Scientific Interest (SSSI), 95, 126, 138–9, 143, 144, 148, 231, 238
Sitka spruce, 226
Skiro, Sweden, 209

Skye, 23, 66, 117, 149, 252
Bridge, 249
Skye and Lochalsh Enterprise, 64
Skye–Lochalsh footpath, 107
Slow Reckoning, 7–8
Small is Beautiful, 6
Smout, T. C., viii, 5, 124, 139–44, 157, 219, 227
Smyth, John, 82, 86–9
Snowsport Scotland, 57
Social Inclusion Research Bulletin, 159
Socialist Party, Scotland, 12
Society of Antiquaries of Scotland, 57
solar energy, 166, 173, 182, 205, 206–7
Spain, 143
Special Areas of Conservation (SAC), 126
Spence, Alan, 24, 162
Speyside, 141
Forest of Spey, 238
Sport 21, 109–11
sports and leisure, 60, 107, 108–114, 116, 126–9, 130–2, 142–4, 146, 244–5, 250
SportScotland, 109, 110–12, 113, 115–16
Standing Conference on Community Development, 62, 66
Starkey, David, 33
State We're In, The, 163
Stirling, 24, 234
Stockholm Technical University, 209
Strathspey, 85
Sudjic, Deyan, 179–80
Sullom Voe, 71
Sustainable Communities Network Scotland, 61, 75n
sustainable development, 8,13, 50–3, 58, 63, 65–6, 111, 125–7, 220–3, 231–9, 240n
and architecture, 1, 166–7, 182
and zoning, 127–8, 142–4
Sustainable Rural Housing in Scotland, 168
Sustainable Scotland, 82–3
sustainable technology, research on, 191–209
Sustrans Scotland, 57
Sutherland, 64, 218–19, 230–1, 237
North Sutherland Community Forestry Trust, 231–3
Sutton, 163–4
Sweden, Swedish, ix, 39, 41–2, 70, 86, 121–3, 132, 134–7, 172, 198–209, 215
forests, 232
funding of environmental research, 202–5
writing, 18, 41–2
Sweden–Denmark Bridge, 203–4
Swedish Academy of Sciences, 201–2, 205
Swedish Department of the Environment, 200
Swedish Eco-tourism Society, 97–8
Swedish Environmental Protection Agency, 121–3, 134–7, 200, 202
Swedish Recycling Association for Packaging, 209
Swift, Jonathan, 19–20
Swindon, Wiltshire, 161
Swing Hammer Swing!, 26

Targeted Inputs for a Better Rural Environment, 138
Tay River, Tayside, 27, 175–6
Native Woods, 234
tenement, concept of, 157, 162–3
Thaw, Duncan, 19
These Demented Lands, 29
'Third Way' , 57, 68
Thistle Award, 104
Thomas, Rachel, 217–18
Thompson, Frank, 176
Thubron, Colin, 20
Tiree, 66, 67–8,
TITAN, 63–4, 75n, 221
Tokyo-Yokohama, 158
Torneträsk Lake, 122
Torrington, Jeff, 26
Toryglen, 72
Tourism and Environment Forum, 106
tourism, 8–9, 11, 13, 53–4, 59–60, 65, 70–1, 96–117, 122, 129, 130–2, 148, 149, 176, 193, 239, 250
see also eco-tourism
Tourist Management Programmes, 105, 106, 107, 131, 132
traffic, transport, 3, 13, 70, 71, 116, 130–2, 185, 196, 203–4, 248
Trainspotting, 26, 35
Tranter, Nigel, 29
Tressour Wood, 169
Trossachs, 105, 106, 124–9, 234, 235
Troubled Waters, 6
Tumulus, 27

UK New Opportunities Fund, 150
'United Kingdom', term, 36
United Nations, 7
Centre for Human Settlements, *see* Habitat II

Conference on Environment and Development (UNCED), 88
Educational, Scientific and Cultural Organisation (UNESCO), 88
urban regeneration, 158–88
and architecture, 163–88
in Glasgow, 176–85
Urban Taskforce, 166
urbanisation, 157–8
USA, 6, 54, 97, 114, 142, 182, 196, 202, 250

Very Quiet Street, A, 29
Vienna, 202
Vink, J., 187,
VisitScotland, 98–100, 117, 210
volatile organic compounds (VOCs), 195–6
Voluntary Action, 57

Wales, 6, 32, 69, 163, 173, 182, 222, 250
Wall, Jeff, ix
Wallace, Jim, 12
Warmcel, 182
Warner, Alan, 29
waste management, 7, 10, 69, 162, 182, 183, 195–6, 249–50
Watson, Alan, 107
Way to Go, 24
Wells, Mark, 89–93
Welsh, Irvine, 26
West Highland Way, **81**
Western Isles, 23, 66, 215
Western Isles Enterprise, 64
Westminster Parliament, 54, 86, 236
Whale and Dolphin Conservation Society, 100, 101
whale watching, 100–2
When they Lay Bare, 20–1
Who Owns Scotland, 145
Who Owns Scotland?,145
Wigan, Michael, 146
Wightman, Andy, 62–3, 145, 146, 149, 150–1
wilderness, concept of, 107, 117, 124
Wildfowl and Wetlands Trust, 57
wildlife charities, 22–3, 141–4
see also Royal Society for the Protection of Birds, World Wide Fund for Nature
wildlife habitats, 95–6, 100–102
Wilson, Norman, 45
wind energy, 73, 197, 222, 223
Wishart, Colin, 167–8, 174–6
Wood, Emma, 36
Wood, Michael, 32
Woodland Grants System, 230, 234
Woodland Trust Scotland, 57
World Business Centre for Sustainable Development, 70
World Conservation Union, 87, 88, 101, 232
World Information Service on Energy (WISE Group), Scotland, 61
World We Have Lost, The, 33
World Wide Fund for Nature, 67, 89
Scotland, 57, 66, 67, 106

Zetland Act, 71–2
Zettersten, Gunnar, 134–7
'zoning', 121–3, 125, 128, 132, 140, 159, 162
Zwarts & Jansma, architects and designers, 207–8